Applying Soft Computing
in Defining Spatial Relations

Studies in Fuzziness and Soft Computing

Editor-in-chief
Prof. Janusz Kacprzyk
Systems Research Institute
Polish Academy of Sciences
ul. Newelska 6
01-447 Warsaw, Poland
E-mail: kacprzyk@ibspan.waw.pl
http://www.springer.de/cgi-bin/search_book.pl?series=2941

Pascal Matsakis
Les M. Sztandera
Editors

Applying
Soft Computing
in Defining
Spatial Relations

With 87 Figures
and 16 Tables

Physica-Verlag

A Springer-Verlag Company

Prof. Pascal Matsakis
University of Missouri
CECS Department, EBW 201
Columbia, MO 65211-2060
USA
pmatsakis@cecs.missouri.edu

Prof. Les M. Sztandera
Philadelphia University
CIS Department
Philadelphia, PA 19144
USA
sztanderal@philau.edu

ISSN 1434-9922
ISBN 978-3-662-00294-0

Library of Congress Cataloging-in-Publication Data applied for
Die Deutsche Bibliothek – CIP-Einheitsaufnahme
Applying soft computing in defining spatial relations: with 16 tables / Pascal Matsakis; Les M. Sztandera,
eds. – Heidelberg; New York: Physica-Verl., 2002
   (Studies in fuzziness and soft computing; Vol. 106)
   ISBN 978-3-662-00294-0   ISBN 978-3-7908-1752-2 (eBook)
   DOI 10.1007/978-3-7908-1752-2

Physica-Verlag Heidelberg New York
a member of BertelsmannSpringer Science+Business Media GmbH

© Physica-Verlag Heidelberg 2002
Softcover reprint of the hardcover 1st edition 2002

Hardcover Design: Erich Kirchner, Heidelberg

SPIN 10880135   88/2202-5 4 3 2 1 0 – Printed on acid-free paper

Foreword

Geometric properties and relations play central roles in the description and processing of spatial data. The properties and relations studied by mathematicians usually have precise definitions, but verbal descriptions often involve imprecisely defined concepts such as elongatedness or proximity. The methods used in soft computing provide a framework for formulating and manipulating such concepts.

The importance of soft concepts in image analysis and computer vision has been recognized for over 30 years. Prewitt pointed out in 1970 [1] that the results of image segmentation should be regarded as fuzzy subsets of the image rather than as "crisp" subsets. The earliest work by Zadeh on fuzzy sets, in 1965 [6], introduced a fuzzification of the concept of convexity. The 1970 Ph.D. dissertation by Winston [5], on learning structural descriptions in the "blocks world," included a chapter on defining relations of relative position (such as "to the left of") between objects.

Since the late 1970's, on the order of 100 papers have appeared on geometric properties of fuzzy sets. Reviews of this work were published in 1984 [2], 1992 [3], and 1998 [4]. Most of it deals with geometric properties: topological, metric, convexity, elongatedness and skeletonization, as well as the use of fuzzy concepts in image segmentation. Only a few references have dealt with the fuzzy (or more generally: soft) specification of geometric relations. The writer hopes that the appearance of this volume, which contains eight papers on the soft definition and manipulation of spatial relations, will lead to increased interest in this important topic.

Azriel Rosenfeld
College Park, MD
January, 2002

References

1. J. M. S. Prewitt. Object Enhancement and Extraction. In: B. S. Lipkin and A. Rosenfeld (Eds.), *Picture Processing and Psychopictorics*, pages 75-149. Academic Press, New York, 1970.
2. A. Rosenfeld. The Fuzzy Geometry of Image Subsets. *Pattern Recognition Letters*, 2:311-317, 1984.
3. A. Rosenfeld. Fuzzy Geometry: An Overview. In *FUZZ-IEEE 1992 (IEEE Int. Conf. on Fuzzy Systems)*, pages 113-117, 1992.
4. A. Rosenfeld. Fuzzy Geometry: An Updated Overview. *Information Sciences*, 110: 127-133, 1998.
5. P. H. Winston. *Learning Structural Descriptions from Examples*. Technical Report 231, MIT Artificial Intelligence Laboratory, 1970.
6. L. Zadeh. Fuzzy Sets. *Information and Control*, 8:338-353, 1965.

Preface

The early roots of soft computing can be traced back to Dr. Lotfi A. Zadeh's book chapter on *soft data analysis* [1] published in 1981. The actual concept of "soft computing," however, was not launched until about 10 years later, when the Berkeley Initiative in Soft Computing (BISC), an industrial liaison program, was established at the University of California, Berkeley.

The main characteristics of soft computing are listed below:

- Capability to approximate various kinds of real-world systems;
- Tolerance for imprecision, partial truth, and uncertainty; and
- Learning from the environment.

These characteristics commonly lead to better rapport with reality, low solution cost, robustness, and tractability. Dr. Zadeh, the father of soft computing, emphasized that soft computing provides a solid foundation for the conception, design, and application of intelligent systems employing its member methodologies *symbiotically* rather than in isolation.

Since the soft computing community is not looking for perfect but competitive solutions, there exists an implicit commitment to benefit from the fusion of various methodologies. Such a fusion can lead to cooperative and complementary combinations of the individual methodologies. Constructive fusion thinking has already been extended beyond the individual SC technologies.

Contained in this edited volume are eight chapters discussing various aspects of soft computing (SC) in the field of spatial relations. The idea is to collect, under one cover, original work by authors who have, in most cases, contributed considerably to that field.

In Chapter 1 (pages 1-16), Guesgen shows that fuzzy logic can be used to handle imprecision in spatial relations and combine qualitative spatial relations with quantitative information. In Chapter 2 (pages 17-39), Clementini argues that the integration of fuzzy knowledge into qualitative models allows more effective spatial reasoning to be performed. Both authors discuss the composition of spatial relations—a common way of reasoning about space.

The following three chapters focus more on how to model different spatial relationships in a unified way and/or in various settings (e.g., quantitative and qualitative). In Chapter 3 (pages 41-62), Sztandera projects spatial fuzzy sets onto orthogonal axes and utilizes a concept of dominance. In Chapter 4 (pages 63-98), Bloch relies on mathematical morphology, whereas in Chapter 5 (pages 99-122) Matsakis proposes the use of force histograms.

The last three chapters are more particularly related to Geographic Information Systems (GIS). Chapter 6 (pages 123-155), by Petry et al., discusses an agent-based framework as a solution for issues related to uncertainty in spatial data. In Chapter 7 (pages 157-178), Robinson shows that spatially explicit ecological modeling is a complex domain rich in the potential for intelligent applications using fuzzy spatial relations. Chapter 8 (pages 179-202), by Zhan, presents a fuzzy set model of approximate linguistic terms used in descriptions of topological relations between geographical regions.

The intended audience of this book includes professionals, researchers and developers of software/hardware tools for the design of soft computing-based systems exploiting spatial relationships, and the entire computational intelligence community. It is expected that the reader is a graduate of electrical engineering, computer engineering, or computer science study program with a modest mathematical background. Our book also forms a good basis for Ph.D. level seminars on spatial relationships.

Les M. Sztandera
Philadelphia University
Philadelphia, Pennsylvania, USA

Pascal Matsakis
University of Missouri-Columbia
Columbia, Missouri, USA

References

1. L. A. Zadeh. Possibility Theory and Soft Data Analysis. In: L. Cobb and R. M. Thrall (Eds.), *Mathematical Frontiers of the Social and Policy Sciences*, pages 69–129, Boulder, CO, Westview Press, 1981.

Table of Contents

Understanding the Spatial Organization
of Image Regions by Means of Force Histograms: 99
A Guided Tour

Pascal Matsakis

A Fuzzy Set Model of Approximate Linguistic Terms in Descriptions of Binary Topological Relations Between Simple Regions

F. Benjamin Zhan

About the Editors   203

Fuzzifying Spatial Relations

Hans W. Guesgen*

Computer Science Department, University of Auckland
Private Bag 92019, Auckland, New Zealand

Abstract. Reasoning about space plays an essential role in many cultures. Not only is space, like time, one of the most fundamental categories of human cognition, but also does it structure all our activities and relationships with the external world. Space serves as the basis for many metaphors, including temporal metaphors. It is inherently more complex than time, because it is multidimensional and epistemologically multiple.

The way humans often deal with space in everyday situations is on a qualitative basis, allowing for imprecision in spatial descriptions when interacting with each other. Instead of using an absolute space (i.e., space viewed as a "container", which exists independently of the objects that are located in it), it seems that they prefer a relative space, which is a construct induced by spatial relations over non-purely spatial entities.

In artificial intelligence, a variety of formalisms have been developed that deal with space on the basis of relations between objects. Although most approaches provide some algorithms to reason about such relations, they usually do not make any attempt to address questions like how to handle imprecision in spatial relations or how to combine qualitative spatial relations with quantitative information. Although these questions seem to be unrelated to each other, we show in this chapter that fuzzy logic can provide an answer to both of them.

1 Motivation

Researchers in artificial intelligence have been arguing successfully that human reasoning about space is of a qualitative nature in most everyday situations and that therefore computer systems should support such a form of reasoning (see, e.g., [20]). For most people, a statement like *The post office is beside the city hall* is more natural than a statement like *The coordinates of the post office can be calculated from those of the city hall by adding the vector* $(-28, 10)$ *to the latter*. It is therefore not surprising that a number of publications in the area of spatial reasoning aim at some form of qualitative reasoning about space. Many of them do so by using a relational approach.

In [14], for example, we introduce a form of spatial reasoning that extends Allen's temporal logic [1] to the three dimensions of space by applying very simple methods for constructing higher-dimensional models and for reasoning about them, namely combination (i.e., building tuples of one-dimensional

* Partly supported by the University of Auckland Research Committee through various grants (most recently through grant number XXX/9343/3414100).

relations) and projection (i.e., extracting one-dimensional aspects from the tuples).

Other approaches proceed in more or less the same way. Freksa [6] uses the same set of relations as in [1]. He shows that for an important class of problems, only a small subset of all possible combinations of spatial relations can occur. By restricting himself to sets of conceptually neighboring relations, he can restrict the complexity of the constraint satisfaction algorithms significantly.

In [20], Hernández introduces an extension of Allen's approach to represent the spatial features occurring in 2D projections of 3D scenes. He suggests to establish spatial relations between objects by splitting them up into two aspects: projection and orientation. The aspect of projection describes the spatial relationship between two objects in a way similar to the one introduced in [14]. The aspect of orientation states how the objects are located relative to each other.

Mukerjee and Joe's work [22] is similar to Hernández's approach. Objects of a two-dimensional world are characterized by the directions in which the objects are moving and by associating with the objects trajectories along which they are moving.

In the context of spatial relations between objects, a reoccurring issue is the choice of an adequate reference frame. In [5], Frank discusses a taxonomy of frames of reference, which is inspired by cognitive models. Since we are not going into details here about choosing an adequate reference frame, we refer the reader to Frank's publication.

A common feature of the approaches sketched above (and many other approaches not referenced here) is that they represent spatial information in the form of qualitative spatial relations between objects:

- *The church is near the post office.*
- *Object O_1 overlaps object O_2.*
- *The pencil is in the drawer.*

Most approaches provide some algorithms to reason about such relations, but they usually do not make any attempt to address the following questions:

- How can imprecision in spatial relations be dealt with?
- How can qualitative spatial relations be combined with quantitative information?

These questions seem to be unrelated to each other, but we show in this chapter that fuzzy logic can provide an answer to both of them.

2   Imprecision in Spatial Relations

2.1   Conceptual Neighborhoods

Allen [1] introduced a temporal logic based on a set of thirteen atomic temporal relations between time intervals (see Figure 1), together with an algorithm

Relation	Illustration	Interpretation
$O_1 < O_2$ $O_2 > O_1$		O_1 before O_2 O_2 after O_1
$O_1 m O_2$ $O_2 mi O_1$		O_1 meets O_2 O_2 met by O_1
$O_1 o O_2$ $O_2 oi O_1$	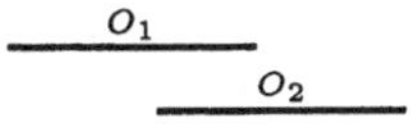	O_1 overlaps O_2 O_2 overlapped by O_1
$O_1 s O_2$ $O_2 si O_1$	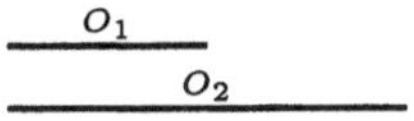	O_1 starts O_2 O_2 started by O_1
$O_1 d O_2$ $O_2 di O_1$	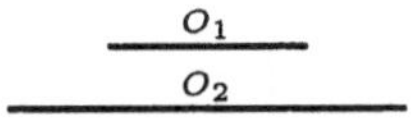	O_1 during O_2 O_2 contains O_1
$O_1 f O_2$ $O_2 fi O_1$	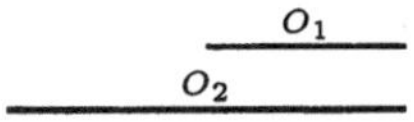	O_1 finishes O_2 O_2 finished by O_1
$O_1 = O_2$	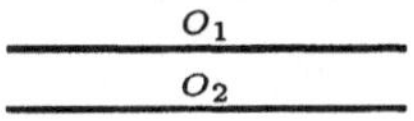	O_1 equals O_2

Fig. 1. Allen's thirteen atomic relations.

to reason about networks of such relations. As indicated in the introductory section of this chapter, Allen's logic can be adopted for spatial reasoning by interpreting the thirteen Allen relations as spatial relations between objects. By applying Allen's algorithm to these spatial relations, we obtain an instrument to reason about space.

The basis of Allen's algorithm is a composition table, which determines the possible relations between two objects like O_1 and O_3 given the relations between O_1 and another object O_2 as well as the relation between O_2 and O_3. The composition table is shown in Figure 2. If, for example, O_1 is enclosing O_2 ($O_1 di O_2$) and O_2 is overlapping O_3 ($O_2 o O_3$), then it can be concluded

	<	m	o	fi	di	si	=	s	d	f	oi	mi	>
<	<	<	<	<	<	<	<	<	<, m o, s d	<, m o, s d	<, m o, s d	<, m o, s d	<, m o, s, d f, =, fi di, si, oi mi, >
m	<	<	<	<	<	m	m	m	o, s d	o, s d	o, s d	f, = fi	di si, oi mi, >
o	<	<	<, m o	<, m o	<, m o fi, di	o fi, di	o	o	o, s d	o, s d	o, s d, f, = fi, di si, oi	di si, oi	di si, oi mi, >
fi	<	m	o	fi	di	di	fi	o	o, s d	f, = fi	di si, oi	di si, oi	di si, oi mi, >
di	<, m o fi, di	o fi, di	o fi, di	di	di	di	di	o fi, di	o, s d, f, = fi, di si, oi	di si, oi	di si, oi	di si, oi	di si, oi mi, >
si	<, m o fi, di	o fi, di	o fi, di	di	di	si	si	s, = si	d, f oi	oi	oi	mi	>
=	<	m	o	fi	di	si	=	s	d	f	oi	mi	>
s	<	<	<, m o	<, m o	<, m o fi, di	s, = si	s	s	d	d	d, f oi	mi	>
d	<	<	<, m o, s d	<, m o, s d	<, m o, s, d f, =, fi di, si, oi mi, >	d, f oi, mi >	d	d	d	d	d, f oi, mi >	>	>
f	<	m	o, s d	f, = fi	di si, oi mi, >	oi mi, >	f	d	d	f	oi mi, >	>	>
oi	<, m o fi, di	o fi, di	o, s d, f, = fi, di si, oi	di si, oi	di si, oi mi, >	oi mi, >	oi	d, f oi	d, f oi	oi	oi mi, >	>	>
mi	<, m o fi, di	s, = si	d, f oi	mi	>	>	mi	d, f oi	d, f oi	mi	>	>	>
>	<, m o, s, d f, =, fi di, si, oi mi, >	d, f oi, mi >	d, f oi, mi >	>	>	>	>	>	d, f oi, mi >	d, f oi, mi >	>	>	>

Fig. 2. Allen's composition table including = as arranged in [7]. The entry at row r_1 and column r_2 in the table denotes the possible relations between O_1 and O_3, assuming that $O_1 r_1 O_2$ and $O_2 r_2 O_3$.

that the relation between O_1 and O_3 is an element of the set $\{o, fi, di\}$. Such a set is called a non-atomic Allen relation.

To adopt the composition table for imprecise reasoning, we explore the notion of conceptional neighbors [7]. Assume that two objects O_1 and O_2, for example, are in relation m, then by moving or deforming the objects slightly we can change this relation to < or o. Therefore, < and o are conceptual neighbors of m. The relation f, for example, is not a conceptual neighbor of m, as f cannot be obtained directly from m by deforming or moving objects.

Freksa [7] distinguishes three different types of conceptual neighbors (A-, B-, and C-neighbors), depending on how the objects can be changed. For example, if the objects can be moved but not deformed, s is *not* a conceptual neighbor of =. However, if we allow for deformation, s and = are neighbors. To visualize the notion of conceptual neighbors, Freksa rearranged the Allen relations in such a way that conceptual neighbors are close in the topological sense if they are close in the conceptual sense. The result is shown in Figure 3 for the case of allowing movement of objects but no deformation. In the

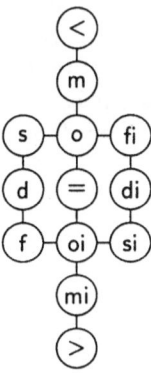

Fig. 3. Allen's thirteen atomic relations arranged according to their conceptual neighborhood. Only movement of objects is considered, no deformation.

following, we restrict our discussions to this case.

2.2 Fuzzification of Allen Relations

The notion of conceptual neighbors can be used to introduce imprecision into reasoning about spatial relations [15]. The first step in that directions involves the introduction of characteristic functions to denote atomic relations:

$$\mu_r : A \longrightarrow \{0, 1\}$$

The domain of μ_r is the set of atomic Allen relations, i.e.:

$$A = \{<, m, o, fi, di, si, =, s, d, f, oi, mi, >\}$$

The function yields a value of 1 if and only if the argument is equal to the atomic relation denoted by the characteristic function:

$$\mu_r(r') = \begin{cases} 1, \text{ if } r' = r \\ 0, \text{ else} \end{cases}$$

The next step towards the introduction of imprecision is to transform the atomic Allen relations into fuzzy sets. For that purpose, we represent each atomic relation as a set of pairs, each pair consisting of an element of $\mathcal{A}$ and the value of the characteristic function of the relation applied to that element. For example, if two object O_1 and O_2 are adjacent to each other, i.e., $O_1 m O_2$, we use the characteristic function of the relation m to convert this statement into the following:

$$O_1\{(r, \mu_m(r)) \mid r \in \mathcal{A}\}O_2 \; = \; O_1\{(m,1),(<,0),(s,0),\ldots\}O_2$$

Instead of having two classes, one with the accepted relations where μ_m results in 1 and another with the discarded relations where μ_m results in 0, we now assign acceptance grades (or membership grades, to use the term from fuzzy set theory) with the relations. If the relation is m, we assign the membership grade 1; if the relation is a neighbor of m, we choose a membership grade α_1 with $1 \geq \alpha_1 \geq 0$; if the relation is a neighbor of a neighbor of m, we assign a grade α_2 with $\alpha_1 \geq \alpha_2 \geq 0$; and so on. Figure 4 illustrates this example.

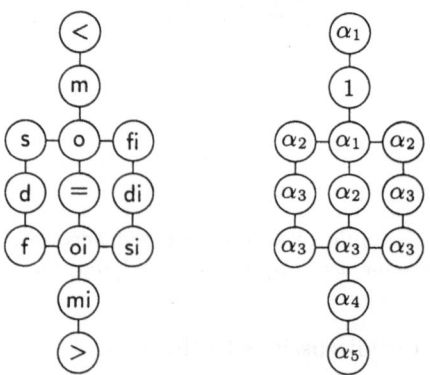

Fig. 4. The atomic Allen relations and their membership grades with respect to the relation m.

Non-atomic Allen relations can be transformed into fuzzy Allen relations by using the same technique. A non-atomic Allen relation is given by a set of atomic Allen relations, which is interpreted in a disjunctive way. We therefore transform each atomic relation in the set into a fuzzy Allen relation and compute the fuzzy union of the resulting sets.

Formally, a fuzzy Allen relation $\tilde{r}$ can be defined by using a function Δ that denotes the conceptional distance between the relation r and a relation r', i.e., Δ results in 1 if r is a neighbor of r', in 2 if r is a neighbor of a neighbor of r', and so on:

$$\Delta : \mathcal{A} \times \mathcal{A} \longrightarrow \{0, 1, 2, \ldots\}$$

Δ can be defined recursively as follows:

1. If $r = r'$, then $\Delta(r, r') = 0$
2. Otherwise, $\Delta(r, r') = \min\{\Delta(r, r'') + 1 \mid r''$ neighbor of $r'\}$

Given a sequence of membership grades, $1 = \alpha_0 \geq \alpha_1 \geq \alpha_2 \geq \cdots \geq 0$, the function Δ can be used to associate Allen relations with membership grades, depending on some given Allen relation r. In particular, we can define a membership function $\mu_{\tilde{r}}$ as follows:

$$\mu_{\tilde{r}} : \mathcal{A} \longrightarrow [0, 1]$$

$$\mu_{\tilde{r}}(r') = \alpha_{\Delta(r,r')}$$

With this definition, the fuzzy Allen relation $\tilde{r}$ of an Allen relation $r \in \mathcal{A}$ is given by the following:

$$\tilde{r} = \{(r', \mu_{\tilde{r}}(r')) \mid r' \in \mathcal{A}\}$$

The values $\alpha_1, \alpha_2, \ldots$, are to be provided by the user. If the user does not feel comfortable with choosing numeric values for the alphas, reasoning can be performed on symbolic values, since it is sufficient to know that there is an order on the alphas. The special case of $\alpha_1 = \alpha_2 = \cdots = 0$ leads to traditional, crisp Allen reasoning.

We now extend the formulation of Allen relations as characteristic functions to the composition of Allen relations, starting again with crisp relations and continuing with fuzzy relations. In the crisp case, Allen's composition table can be represented as a set of characteristic functions of the following form:

$$\mu_{r_1 \circ r_2} : \mathcal{A} \longrightarrow \{0, 1\}$$

The domain of $\mu_{r_1 \circ r_2}$ is the set of atomic Allen relations, i.e.:

$$\mathcal{A} = \{<, \mathsf{m}, \mathsf{o}, \mathsf{fi}, \mathsf{di}, \mathsf{si}, =, \mathsf{s}, \mathsf{d}, \mathsf{f}, \mathsf{oi}, \mathsf{mi}, >\}$$

The function yields a value of 1 for arguments that are elements of the corresponding entry in the composition table (row $r_1 \times$ column r_2, denoted by $r_1 \circ r_2$); otherwise, a value of 0:

$$\mu_{r_1 \circ r_2}(r) = \begin{cases} 1, & \text{if } r \subseteq r_1 \circ r_2 \\ 0, & \text{else} \end{cases}$$

For example, if $r_1 = $ m and $r_2 = $ di, then the characteristic function of the relation $r_1 \text{o} r_2 = $ modi is defined as follows:

$$\mu_{\text{modi}}(r) = \begin{cases} 1, \text{ if } r = < \\ 0, \text{ else} \end{cases}$$

Adopting the min/max combination scheme from fuzzy set theory, we can now define the fuzzy composition $\tilde{r}_1 \text{o} \tilde{r}_2$ of two fuzzy Allen relations $\tilde{r}_1$ and $\tilde{r}_2$ as the fuzzy Allen relation $\{(r, \mu_{\tilde{r}_1 \text{o} \tilde{r}_2}(r)) \mid r \in \mathcal{A}\}$, where $\mu_{\tilde{r}_1 \text{o} \tilde{r}_2}$ is given by the following:

$$\mu_{\tilde{r}_1 \text{o} \tilde{r}_2}(r) = \max_{r'_1, r'_2 \in \mathcal{A} \mid \mu_{r'_1 \text{o} r'_2}(r) = 1} \{\min\{\mu_{\tilde{r}_1}(r'_1), \mu_{\tilde{r}_2}(r'_2)\}\}$$

For example, to compute the fuzzy composition $\tilde{\text{mo}}\text{di}$, we have to determine $\mu_{\tilde{\text{m}}}$ and $\mu_{\tilde{\text{di}}}$. The resulting membership grades have to be combined according to the above min/max scheme. We still apply the original Allen table to determine the arguments for $\mu_{\tilde{\text{m}}}$ and $\mu_{\tilde{\text{di}}}$ and consider only those $r'_1, r'_2 \in \mathcal{A}$ whose combination is a set of Allen relations containing r, i.e., the Allen relation for which we want to compute the membership grade. Figure 5 shows a graphical representation of $\tilde{\text{mo}}\text{di}$.

3   Applying Allen's Algorithm to Fuzzy Relations

Input to Allen's algorithm is a set of objects and a set of (not necessarily atomic) Allen relations. If there is no relation specified for a particular pair of objects, it is assumed that the relation between these objects is the set of all thirteen atomic Allen relations. The aim of the algorithm is to transform the given relations into a set of relations that are consistent with each other. This is achieved through an iterative process that looks at three objects at a time and applies the composition table to these objects. Any relation that is not justified by the composition table (i.e., which is not listed in the table) is inconsistent with the rest of the given relations and therefore is deleted by the algorithm.

For example, let us assume that the initial relations between the objects O_1, O_2, and O_3 are $O_1 \text{s} O_2$, $O_2 \text{o} O_3$, and $O_1\{\text{m}, \text{o}, \text{fi}, \text{di}\}O_3$. A look-up of the relation between O_1 and O_3 in the composition table (i.e., the relation at row s and column o) yields $\{<, \text{m}, \text{o}\}$, which means that fi and di have to be deleted from the initial relation between O_1 and O_3, resulting in the new relation $O_1\{\text{m}, \text{o}\}O_3$. In more general terms, this step can be formulated as follows:[1]

$$r(O_1, O_3) \longleftarrow r(O_1, O_3) \cap [r(O_1, O_2) \text{o} r(O_2, O_3)]$$

The extension of Allen's algorithm to cope with fuzzy relations is straightforward. Rather than making a yes/no decision about whether a relation is

---

[1] $r(O_i, O_j)$ denotes the relation between O_i and O_j for $i, j \in \{1, 2, 3, \ldots\}$.

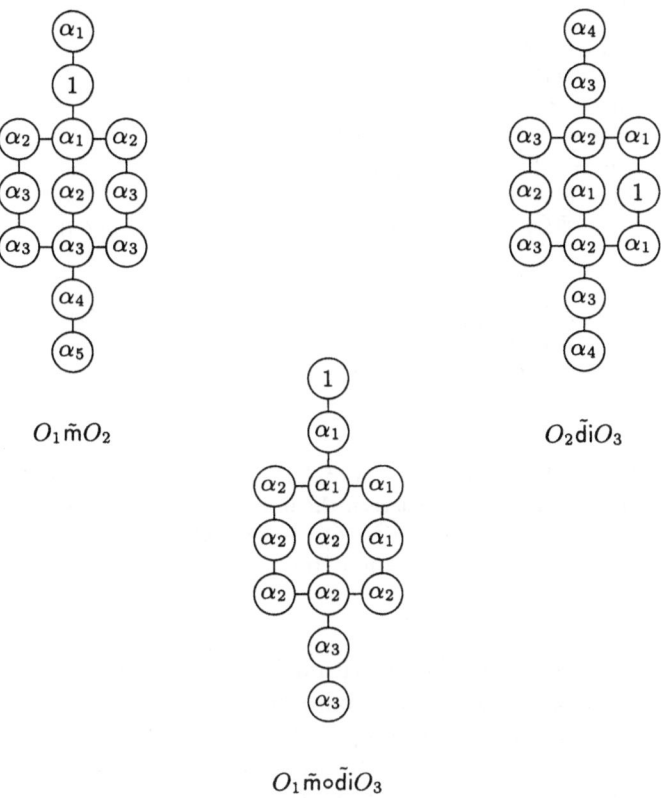

Fig. 5. The fuzzy composition of m and di.

an element of the composition of two other relations, we compute a membership grade for that relation. This membership grade is compared with the initial membership grade of the relation. If the new grade is smaller than the initial grade, the membership grade of the relation is updated with the new grade. Analogously to non-fuzzy Allen relations, this step can be formulated as follows:

$$\tilde{r}(O_1, O_3) \longleftarrow \tilde{r}(O_1, O_3) \cap [\tilde{r}(O_1, O_2) \circ \tilde{r}(O_2, O_3)]$$

The intersection of two fuzzy Allen relations $\tilde{r}(O_1, O_3)$ and $\tilde{r}'(O_1, O_3)$ is defined in the usual way of minimizing membership grades:

$$\tilde{r}(O_1, O_3) \cap \tilde{r}'(O_1, O_3) = \{(r, \min\{\mu_{\tilde{r}(O_1,O_3)}(r), \mu_{\tilde{r}'(O_1,O_3)}(r)\}) \mid r \in \mathcal{A}\}$$

Figure 6 shows pseudocode for the extended algorithm; a more elaborate discussion of the algorithm can be found elsewhere [17]. Unlike in the crisp version of Allen's algorithm, no elements are deleted from the fuzzy Allen

Given a set of objects $\{O_1, O_2, \ldots, O_n\}$
Given a set $\tilde{\mathcal{R}}$ of fuzzy Allen relations between these objects.
While $\tilde{\mathcal{R}}$ is not empty:
    Select a relation $\tilde{r}(O_i, O_j) \in \tilde{\mathcal{R}}$
    $\tilde{\mathcal{R}} \longleftarrow \tilde{\mathcal{R}} - \{\tilde{r}(O_i, O_j)\}$
    For $k \in \{1, \ldots, n\}$ with $k \neq i, j$:
        $\tilde{r}(O_k, O_j) \longleftarrow \tilde{r}(O_k, O_j) \cap [\tilde{r}(O_k, O_i) \circ \tilde{r}(O_i, O_j)]$
        If $\tilde{r}(O_k, O_j)$ changed:
            $\tilde{\mathcal{R}} \longleftarrow \tilde{\mathcal{R}} \cup \{\tilde{r}(O_k, O_j)\}$
        $\tilde{r}(O_i, O_k) \longleftarrow \tilde{r}(O_i, O_k) \cap [\tilde{r}(O_i, O_j) \circ \tilde{r}(O_j, O_k)]$
        If $\tilde{r}(O_i, O_k)$ changed:
            $\tilde{\mathcal{R}} \longleftarrow \tilde{\mathcal{R}} \cup \{\tilde{r}(O_i, O_k)\}$

Fig. 6. Pseudo-code for a fuzzy version of Allen's algorithm. A term like $\tilde{r}(O_i, O_j)$ denotes the fuzzy Allen relation between the objects that it is referring to. Without loss of generality, we assume that $\tilde{r}(O_i, O_j)$ is defined for every $i, j \in \{1, 2, \ldots, n\}$ with $i \neq j$, possibly as universal relation $\{(\mathsf{m}, 1), (<, 1), (\mathsf{s}, 1), \ldots\}$.

relations during the run of the algorithm (but their membership grades are updated). This means that in each composition step, the 13 elements of one fuzzy Allen relation have to be composed with the 13 elements of the other fuzzy Allen relation, resulting in 169 table look-ups. These table look-ups result in a total of 334 atomic relations (as some table entries contain more than one relation), each of which is associated with the minimum of the membership grades of the relations that led to this relation. From these 334 atomic relations, the elements are selected whose membership grades are maximal with respect to $<$, m, o, and so on.

In order to avoid extensive, often redundant, search for the best relation, two different strategies can be exploited. The first strategy avoids extensive search by filling relations on a best-first basis instead of determining the membership grade of each element in the new fuzzy Allen relation in turn by first minimizing the pairs and then maximizing results. This method results from the following considerations:

1. A membership grade of 1 in the composed relation can only result from combining relations with membership grades of 1 in the original two relations.

2. Assuming that all such membership grades have been filled in, membership grades of α_1 can only result from combining original relations which have membership grades of 1 or α_1, and consequently membership grades of α_2 from original membership grades of 1, α_1, of α_2, etc.

Search is therefore able to stop as soon as a membership grade has been obtained for each of the 13 Allen relations, because, by virtue of the heuristic, the first value obtained must be the maximum.

The second strategy addresses the problem of repeated look-ups. During the composition of two fuzzy Allen relations, the same look-up pair of atomic Allen relations is often produced several times. To avoid that a combination of relations is looked up more than once, a hash table is maintained in which pairs are recorded that have already been looked up. Before any two relations are composed, this hash table can be consulted to ensure an equivalent combination has not already been processed.

4   Other Fuzzy Relations

So far, we have restricted ourselves to a particular type of relation, namely Allen relations. We show in the following that other object relations, such as proximity relations, can be dealt with in a similar way. This leads us to a more general framework for qualitative reasoning about spatial information [13].

Consider, for example, a geographic information system and the problem of finding a suitable location for a new city dump given certain constraining factors such as the following:

1. The location must be within 500 meters of an existing road.
2. The location must have an area of more than 1000 square meters.
3. The location must be at least 1000 meters from residential or commercial property.
4. The location must be at least 500 meters from any water.
5. The location must not be situated on land covered in native vegetation.

It is possible that a search observing these factors eventually fails, as they restrict the search space too dramatically by excluding any locations, for example, which are 490 meters from water or 510 meters from a road. Such a location, however, might be the best choice available and therefore perfectly acceptable.

The problem of over-constrained search can be solved by using qualitative spatial statements rather than quantitative ones. Instead of *All locations that are within 500 meters of a road*, we would put in the restriction *All locations that are close to a road*. The system would then analyze the qualitative relation *close to a road* that is used in this restriction and would find the areas that best match this restriction and that are compatible with the other restrictions.

One way of analyzing qualitative relations and finding a best match is to associate each qualitative relation with a fuzzy set and apply fuzzy logic to these sets. The first step in this approach is to interpret qualitative spatial relations among objects as restrictions of spatial linguistic variables. Informally, a linguistic variable is a variable whose values are words or phrases in a natural or artificial language, like *close to a road* or *away from water* [27]. The values of a linguistic variable are called linguistic values. A restriction of a linguistic variable is a subset of its linguistic values. For example, the position of the city dump can be represented by a linguistic variable whose linguistic values are restricted to the domain {*close to a road*}.[2]

To be able to reason about restrictions of linguistic variables, we associate a fuzzy (sub)set with each linguistic variable. This fuzzy set is the union of the fuzzy sets that correspond to the linguistic values of the variable. Each linguistic value is associated with exactly one fuzzy set. For example, the value *close to a road* may be associated with a fuzzy set that characterizes for each coordinate on some given map to which extent this coordinate represents some location close to a road. Assuming that the underlying domain for this fuzzy set, $D = \{A1, A2, A3, \ldots, B1, B2, \ldots, Z1, \ldots\}$, represents the possible coordinates as character–digit combinations, *close to a road* may be associated with the following fuzzy set:

$$\tilde{R} = \{\langle M5, 1\rangle, \langle M4, 0.8\rangle, \langle M6, 0.8\rangle, \langle L5, 0.8\rangle, \langle N5, 0.8\rangle, \langle L4, 0.7\rangle, \ldots\}$$

In other words, each location is considered to be more or less close to a road. If its membership grade equals 1, the location is definitely close to a road; if it equals 0, then it is not close to any road at all.

The association of linguistic variables with fuzzy sets provides us with a means to reason about linguistic values. The idea is to view restrictions of linguistic variables as fuzzy constraints and to apply constraint satisfaction algorithms to the resulting fuzzy constraint networks. This approach is closely related to applying Allen's algorithm to fuzzy Allen relations, as discussed in the previous section, since Allen's algorithm can be viewed as a particular kind of constraint satisfaction algorithm.

5   Fuzzy Constraint Satisfaction

A constraint network consists of a set of variables and a set of relations among these variables, called constraints [21]. In a fuzzy constraint network, each constraint is associated with some extra information. Given an assignment of values to the variables of the constraint, this information indicates how well the constraint is satisfied by the value assignment [11]. For example, assigning a value of *M4* to a variable that is constrained by a *close to a road* constraint would result in a satisfaction grade of 0.8 (assuming that the constraint is defined as indicated in the previous example).

---

[2] In this example, the restriction is applied to a single linguistic variable (rather than a pair, triplet, etc. of linguistic variables) and it contains only one value. This in not the case in general.

In other words, a fuzzy constraint network consists of a set of linguistic variables and a set of fuzzy constraints $\tilde{R}_1, \ldots, \tilde{R}_n$, each $\tilde{R}_i$ $(i = 1, \ldots, n)$ ranging over a subset of the variables. The main operations performed on the constraints of the constraint network are union and intersection. A constraint may be viewed as union of one-element sets, each of which represents a possible choice of values for the variables of the constraint, i.e., a combination of locations that satisfies the constraint. A constraint network, on the other hand, may be viewed as intersection of the relations represented by the constraints of the network.

Usually, we are not interested in computing the entire relation represented by a fuzzy constraint network, but want to obtain an element of this relation whose membership grade is beyond a certain threshold α. Such an element is called an α-solution of the fuzzy constraint network.

An algorithm often used for finding an α-solution of a fuzzy constraint network is branch and bound [10,18,24], which operates in the same way as backtracking with two variations:

1. The best solution so far is recorded.
2. A search path is abandoned when it becomes clear that it cannot lead to a better solution.

Search stops when all search paths have been either explored or abandoned, or when a perfect solution has been found. In the case of fuzzy constraint satisfaction, a perfect solution would be a 1-solution.

Since pure branch and bound is not a tractable approach to solving fuzzy constraint satisfaction problems, extra methods are required. They can generally be divided into two categories:

1. Consistency propagation algorithms, which achieve various levels of local consistency.
2. Constructive heuristic search methods, which make choices about forward or backward moves, avoid redundant checking, or order variables or values in a way that might expedite search.

Most of these methods have their roots in the area of traditional constraint satisfaction. For example, the fuzzy consistency propagation algorithms are based on the algorithms described in [21], whereas variable and value ordering heuristics are closely related to the work discussed in [23].

A particularly successful consistency propagation algorithm is forward checking [19], which performs a consistency check each time a variable is instantiated and removes inconsistent values from the domains of the uninstantiated variables. Most empirical studies of constructive heuristic search in constraint satisfaction credit this heuristic with being the most effective [9,23]. Forward checking can be applied equally well to fuzzy constraint networks, but instead of deleting inconsistent values from the variable domains, the algorithm alters the membership grades of the values in these domains.

Whenever a membership grade is reduced to a grade less than or equal to the given threshold α, the corresponding value is discarded, as it cannot participate in an improved solution.

Forward checking can be taken a step further by applying it iteratively. The result is a constraint network which is arc consistent, i.e., in which the domains of each pair of variables are consistent with the constraint between the variables. Mackworth [21] proposed several algorithms to transform constraint networks into arc consistent constraint networks. Dubois et al. [4] adapted one of these algorithms to fuzzy constraint networks.

Although consistency propagation plays an essential role in improving the behavior of branch and bound, it is not the only technique commonly used in constraint satisfaction. It is generally accepted that the order in which variables are instantiated can have a tremendous impact on the size of the search space a backtrack search explores. The problem of finding a variable ordering that minimizes the search space is very difficult, so most research in this area has been aimed at developing heuristics which reduce the search space.

There is a considerable amount of research on variable ordering heuristics for classical constraint satisfaction problems [2,8,19,25]. Many of these are based on the general idea of instantiating the most difficult or constrained variables first, which is justified by the fact that searching first in the most difficult parts of the search space helps to make the failures appear early in the search. This idea transfers directly to fuzzy constraint networks [12].

Considerably fewer heuristics have been applied to the task of ordering the values available for selection within the domain of a variable. Dechter and Pearl [3] suggest this is partly because if a backtrack search for all solutions is being performed, the search tree produced is invariant on the value selection. They point out, however, that the situation differs considerably if only one solution is required. In this case the ordering in which values are selected can have a profound effect on the performance of the algorithm. Dechter and Pearl tested the effects of using different levels of information to order the values for selection. They found out that more benefit was obtained from a fairly weak level of look-ahead.

Value ordering in a fuzzy domain may not necessarily have the same aims as value ordering in a crisp one. It is true that maximizing future options is still a worthwhile goal but it is important to remember that fuzzy constraint networks represent optimization problems. Unlike the values in crisp domains, the fuzzy values that are to be ordered already have in their membership grade a metric indicating their suitability. In solving a fuzzy constraint network, the sooner a solution with a consistency close to optimality is found, the more the search space is able to be pruned. This means that values are to be ordered, either statically or dynamically, according to some measure of their contribution to the optimality of a solution [12].

6   Conclusion

In this chapter, we showed how to fuzzify spatial relations, starting with the set of Allen relations and continuing with general spatial relations represented as constraints. We introduced several methods for reasoning about fuzzy spatial relations, including an extension of Allen's algorithm and methods for fuzzy constraint satisfaction.

This chapter does not claim to be comprehensive. Firstly, there are various approaches to spatial representation and reasoning that have not been addressed here [26]. Secondly, some spatial relations might have to be dealt with in a different way than described in this paper. An example is the notion of proximity and its application in the buffer operations of fuzzy geographic information systems [16].

Acknowledgement

This chapter is based on previously-published joint research with Joachim Hertzberg, Jonathan Histed, Anne Philpott, David Poon, and the author. We are grateful for the support that we received from the University of Auckland Research Committee under various grants.

References

1. Allen J.F. (1983), Maintaining knowledge about temporal intervals, Communications of the ACM, 26, pp. 832–843.
2. Dechter R. and Meiri I. (1994), Experimental evaluation of preprocessing algorithms for constraint satisfaction problems, Artificial Intelligence, 68, pp. 211–241.
3. Dechter R. and Pearl J. (1987), Network-based heuristics for constraint-satisfaction problems, Artificial Intelligence, 34, pp. 1–38.
4. Dubois D., Fargier H., and Prade H. (1993), Propagation et satisfaction de constraintes flexibles, in Fuzzy Sets, Neural Networks and Soft Computing, Yager R.R. and Zadeh L. (Eds.), Kluwer, Dordrecht, The Netherlands.
5. Frank A.U. (1998), Formal models for cognition: taxonomy of spatial location description and frames of reference, in Spatial Cognition: An Interdisciplinary Approach to Representation and Processing Spatial Knowledge, Freksa C., Habel C., and Wender K.F. (Eds.), Lecture Notes in Artificial Intelligence 1404, Springer, Berlin, Germany, pp. 293–312.
6. Freksa C. (1990), Qualitative spatial reasoning, Proc. Workshop RAUM, Koblenz, Germany, pp. 21–36.
7. Freksa C. (1992), Temporal reasoning based on semi-intervals, Artificial Intelligence, 54, pp. 199–227.
8. Freuder E.C. (1982), A sufficient condition for backtrack-free search, Journal of the ACM, 29, pp. 24–32.

9. Freuder E.C. (1994), Using metalevel constraint knowledge to reduce constraint checking, Proc. ECAI-94 Workshop on Constraint Processing, Amsterdam, The Netherlands, pp. 27–33.
10. Freuder E.C. and Wallace R.J. (1992), Partial constraint satisfaction. Artificial Intelligence, 58, pp. 21–70.
11. Guesgen H.W. (1994), A formal framework for weak constraint satisfaction based on fuzzy sets, Proc. ANZIIS-94, Brisbane, Australia, pp. 199–203.
12. Guesgen H.W. (1996), Attacking the complexity of fuzzy constraint satisfaction problems, Proc. International Discourse on Fuzzy Logic and the Management of Complexity (FLAMOC-96), Sydney, Australia, pp. 66–72.
13. Guesgen H.W. and Albrecht J. (2000), Imprecise reasoning in geographic information systems, Fuzzy Sets and Systems (Special Issue on Uncertainty Management in Spatial Data and GIS), 113, pp. 121–131.
14. Guesgen H.W. and Hertzberg J. (1993), A constraint-based approach to spatiotemporal reasoning, Applied Intelligence (Special Issue on Applications of Temporal Models), 3, pp. 71–90.
15. Guesgen H.W. and Hertzberg J. (1996), Spatial persistence, Applied Intelligence (Special Issue on Spatial and Temporal Reasoning), 6, pp. 11–28.
16. Guesgen H.W. and Hertzberg J. (2001), Algorithms for buffering fuzzy raster maps, Proc. FLAIRS-01, Key West, Florida, pp. 542–546.
17. Guesgen H.W., Hertzberg J., and Philpott A. (1994), Towards implementing fuzzy Allen relations, Proc. ECAI-94 Workshop on Spatial and Temporal Reasoning, Amsterdam, The Netherlands, pp. 49–55.
18. Guesgen H.W. and Philpott A. (1995), Heuristics for solving fuzzy constraint satisfaction problems, Proc. ANNES-95, Dunedin, New Zealand, pp. 132–135.
19. Haralick R.M. and Elliott G.L. (1980), Increasing tree search efficiency for constraint satisfaction problems, Artificial Intelligence, 14, pp. 263–313.
20. Hernández D. (1991), Relative representation of spatial knowledge: the 2-D case, in Cognitive and Linguistic Aspects of Geographic Space, Mark D.M. and Frank A.U. (Eds.), Kluwer, Dordrecht, The Netherlands, pp. 373–385.
21. Mackworth A.K. (1977), Consistency in networks of relations, Artificial Intelligence, 8, pp. 99–118.
22. Mukerjee A. and Joe G. (1990), A qualitative model for space, Proc. AAAI-90, Boston, Massachusetts, pp. 721–727.
23. Prosser P. (1993), Hybrid algorithms for the constraint satisfaction problem, Computational Intelligence, 9, pp. 268–299.
24. Ruttkay Z. (1994), Fuzzy constraint satisfaction, Proc. FUZZ-IEEE'94, Orlando, Florida.
25. Stone H.S. and Stone J.M. (1986), Efficient search techniques: an empirical study of the n-queens problem, Technical Report RC 12057 (#54343), IBM T.J. Watson Research Center, Yorktown Heights, New York.
26. Vieu L. (1997), Spatial representation and reasoning in artificial intelligence, in Spatial and Temporal Reasoning, Stock O. (Ed.), Kluwer, Dordrecht, The Netherlands, pp. 5–41.
27. Zadeh L.A. (1975), The concept of a linguistic variable and its application to approximate reasoning—I, Information Sciences, 8, pp. 199–249.

Path Composition of Positional Relations Integrating Qualitative and Fuzzy Knowledge

Eliseo Clementini

Dipartimento di Ingegneria Elettrica
Università di L'Aquila
I-67040 Poggio di Roio, Italy
e-mail: eliseo@ing.univaq.it

Abstract. Qualitative models of spatial knowledge concern the description of both objects and their relative position in space. In this context, fuzzy knowledge concerns the degree of likelihood with which a qualitative spatial relation can be mapped to a geometric counterpart. In this chapter, we argue that the integration of fuzzy knowledge into qualitative models allows us more effective spatial reasoning. In fact, the basic step of reasoning with positional relations, that is, the composition of two relations, if iterated over a path of several intermediate positions, introduces too much indeterminacy in the result. If the algorithms for composition take into account fuzzy knowledge, the latter effect is considerably reduced, obtaining an indication of the degree of likelihood of the result.

1 Introduction

Traditionally, space representation in computer vision, pattern recognition and image analysis is based on the use of quantitative methods. In recent years, qualitative spatial reasoning, a subfield of Artificial Intelligence (AI), has been developed with the aim of modeling commonsense knowledge of space [5], with potential applications in areas as diverse as Geographical Information Systems (GISs), Computer Aided Design (CAD), and Document Recognition. In these areas, the concept of space varies according to the scale, from a table-top environment to the largest geographic space, but it maintains its main characteristics.

The popularity of these methods is due to various factors, including: (a) high-precision quantitative measurements are not as universally useful for the analysis of complex systems as was believed in the beginning of the computer age, when great advances in the natural sciences and in engineering were made as a result of the availability of high-precision measurement and computation equipment; (b) much experience and confidence have been gained in AI regarding the representation and processing of non-numerical knowledge; (c) although storage capacity is much less an issue today than it used to be, it has been recognized that decision making on the basis of large amounts of quantitative knowledge can be computationally expensive; as a consequence, a reduction of data without loss of information remains an important goal;

(d) it has become apparent that higher cognitive mechanisms employ qualitative rather than quantitative mechanisms even if the knowledge originally is available in quantitative form through perception.

Qualitative reasoning can be seen as modeling and simulation with incomplete knowledge. A qualitative representation is a discretisation that makes only those distinctions that are relevant for the global aspects of the domain to be modeled. Qualitative reasoning has been studied extensively for scalar values, such as the temporal domain [1]. Methods and techniques for multidimensional spaces have been developed more recently and still there are many open research issues. The delay in the development of qualitative models and reasoning techniques for spatial domains was partly due to the convincement that "it seems unlikely that such inference schemes will be useful for tasks that require full higher-dimensional manipulations" [10, p. 427]. Recent work in qualitative spatial representations has made more evident that the application of quantity spaces in more than one dimension can lead to promising results. In fact, several aspects of spatial information are currently being investigated [6]. Overall, spatial properties that can be represented in qualitative models can be structured along the three orthogonal axes of topological, projective and metric properties [3].

Positional information, which is taken into account in this chapter, is part of the metric axis in the case of large scale when the distances among objects embedded in space are much bigger than the dimensions of the objects themselves and, therefore, objects can be modeled as zero-dimensional points [7]. In the case of small scale, generally the dimensions of the objects cannot be disregarded and, therefore, positional information must be represented making use also of topological and projective properties. Qualitative distinctions for positional information are structured along various levels of granularity [13]. There are coarser levels and finer levels, each level being independent of the others and characterized by partitioning the configuration space in smaller classes when granularity increases.

Positional information is knowledge made up of orientation and distance relations, which depending on scale can be egocentric (centered on observer), allocentric (relative to distinguished reference structures), or geocentric (relative to coordinated system of reference frames) [12]. Other factors on which both orientation and distance might depend, like the objects' sizes, the scale and the point of view, suggest that it is not enough to express them in terms of isolated relations. Frames of reference are introduced to take into account those internal and external factors. Furthermore, cognitive studies show the existence of multiple frames of reference [17]. For example, when someone emerges from a subway system or a driver gets off a highway to enter a local street network, there is a sudden change in the frame of reference.

Various frameworks for handling qualitative positional relations have been developed in the literature [20,4,16,14]. In all approaches, the basic inference mechanism is the *composition* of spatial relations, that is, given the two relations 'A, r_1, B' and 'B, r_2, C', the problem is to find the relation 'A, r_3, C'. Usually, the composition of relations is defined by an exhaustive table of all possible relation pairs or by "ad hoc" algorithms (as those given in [4]).

In this chapter, we consider the problem of the iteration of the composition of positional relations over a two-dimensional path made up of several intermediate steps. What makes this problem non-trivial is the fact that composition of positional relations does not produce unique results, but a range of possible values for the resulting positional relation. As a consequence, the iteration of composition along a path rapidly leads to a proliferation of possible solutions and, therefore, to the absence of significant answers. The problem of *path composition* has never been addressed before, with the exception of some ideas coming from [21]. We propose an approach to reduce the indeterminacy of results that is based on the combination of qualitative relations with *fuzzy knowledge*. The latter one may come from different sources and, in particular, may intrinsically come from the structure of geometric intervals which qualitative distances can be mapped to. In this manner, the framework for reasoning with qualitative positional relations gains flexibility and becomes more applicable to real cases. Early approaches on combining fuzzy and qualitative techniques for reasoning about distances are notably [9,2,15]. Recently, in [11], the authors studied fuzzy distance and proximity measures as a means to map between quantitative and qualitative representations.

In Section 2, we recall the framework for the qualitative representation of distance and the algorithms for the composition of positional relations originally developed in [4]. In Section 3, we formulate the problem of path composition with some general strategies. In Section 4, we extend the framework with fuzzy knowledge and discuss the composition of so-called fuzzy qualitative positional relations. In Section 5, we apply the path composition of fuzzy qualitative positional relations to a concrete example where additional fuzzy knowledge can be inferred from the knowledge about ratios between intervals of a distance system. In Section 6, we draw short conclusions.

2   Composition of Positional Relations

2.1   Qualitative Distance Relations

In [4], qualitative distance relations have been formalized as follows. Three elements are needed to establish a distance relation: the *primary object* (PO), the *reference object* (RO), and the *frame of reference* (FofR). Let us consider a level of granularity with $n+1$ distance distinctions that partition the space surrounding a reference object RO, and let us name them with a finite set of distance symbols $Q = \{q_0, q_1, q_2, \ldots, q_n\}$, where q_0 is the distance closest to RO and q_n is the one farthest away (see also Figure 1). Given a set of objects O, the qualitative distance between a PO and a RO, both belonging to O, is a function $d : O \times O \to Q$, which associates to the PO the distance symbol identifying the qualitative distance from the RO. If an object A acts as the RO and an object B acts as the PO, the distance between A and B is expressed by $d_{AB} = d(A, B)$. Note that whereas a subscripted d_{AB} denotes a distance *variable* (i.e., the distance of primary object B from a given reference object A), q_i denotes a qualitative distance *value*.

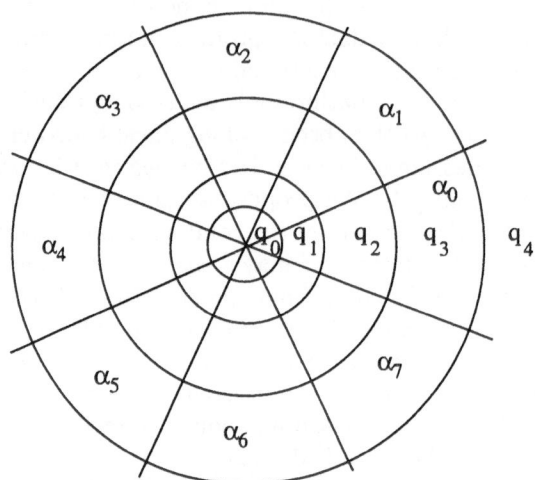

Fig. 1. A qualitative system with 5 distance distinctions $q_0, \ldots, q_4$ and 8 orientation distinctions $\alpha_0, \ldots, \alpha_7$.

Besides naming distances, we also need to specify how they relate to each other, i.e., compare their magnitudes. To this end, we consider a mapping from the distance symbols to 1-dimensional geometric intervals representing distance ranges. Then, an algebraic structure over intervals with order relations is introduced with the purpose of comparing distance ranges. These comparative relations, which should express among others also order-of-magnitude relations, are called structure relations. The idea of mapping qualitative relations to geometric intervals comes from [18].

The three notions mentioned above are organized in *distance systems*. Formally, a distance system D is defined as:

$$D = (Q, \mathcal{A}, \mathcal{I})$$

where:

- Q is the totally ordered set of distance relations;
- $\mathcal{A}$ is an *acceptance function* defined as $\mathcal{A} : Q \times O \rightarrow I$, such that, given a reference object RO, $\mathcal{A}(q_i, RO)$ returns the geometric interval $\delta_i \in I$ corresponding to the distance relation q_i;
- $\mathcal{I}$ is an algebraic structure with operations and order relations defined over a set of intervals I. $\mathcal{I}$ defines the *structure relations* between intervals.

Each distance relation can be associated to an *acceptance area* surrounding a reference object. In the case of isotropic space, acceptance areas are circular regions which can be uniquely identified with a series of consecutive intervals δ_0, δ_1, ..., δ_n (distance ranges). The acceptance function $\mathcal{A}$ performs such a mapping from the symbolic domain of distance relations to geometric intervals. This mapping is necessary to calculate the composition of distances in the domain of intervals, rather than in the domain of distance relations. Then, the inverse function $\mathcal{A}' : I \times O \rightarrow Q$ is used to find the corresponding result back to the domain of distance relations; overall:

$$
\begin{array}{ccc}
Q & \xrightarrow{\ \oplus\ } & Q \times Q \rightarrow Q \\
\downarrow{\scriptstyle\mathcal{A}} & & \uparrow{\scriptstyle\mathcal{A}'} \\
I & \xrightarrow{\ +\ } & I \times I \rightarrow I
\end{array}
$$

The composition of distances in the domain of distance relations is indicated with $\oplus$.

2.2   Composition of Positional Relations

In [4], we described the composition of distance and orientation relations, as the basic step of qualitative reasoning. Given the position of an object B with respect to an object A in terms of qualitative distance and orientation, and the position of a third object C with respect to B, what we want to infer is the position of C with respect to A. We first discussed composition assuming that the frames of reference for orientation and distance are the same in A and B, when they act as reference objects.

Unlike the quantitative sum of vectors, the composition of positional relations cannot be expressed as a formula to compute the resulting position, since angles are only available as orientations and lengths are only available as distance symbols. We cannot always find a unique result for the composition of qualitative distances and orientations, but rather a logical disjunction of possible results. However, since the disjunctive result must be made up of consecutive distances, we proceed by finding a lower and an upper bound for the resulting distances.

Formally, the *position* of an object B with respect to an object A is represented by the pair (d_{AB}, θ_{AB}). Given three objects A, B, and C, if we know the two pairs (d_{AB}, θ_{AB}) and (d_{BC}, θ_{BC}), their *composition* is the pair (d_{AC}, θ_{AC}).

In the following, we investigate how to compute the composition (d_{AC}, θ_{AC}). To that end, we refer to the three basic cases of same $(\theta_{BC} = \theta_{AB})$, opposite $(\theta_{BC} = \text{opp}(\theta_{AB}))$, and orthogonal $(\theta_{BC} \in \text{orth}(\theta_{AB}))$ orientation.

Same Orientation. Let us first suppose that the orientation of B with respect to A is the same as the orientation of C with respect to B: $\theta_{BC} = \theta_{AB}$. As said before, we cannot always find a unique result for the composition of qualitative distances, but rather a logical disjunction of possible results. However, since the disjunctive result must be made up of consecutive distances, we proceed by finding a lower and an upper bound for the resulting distances.

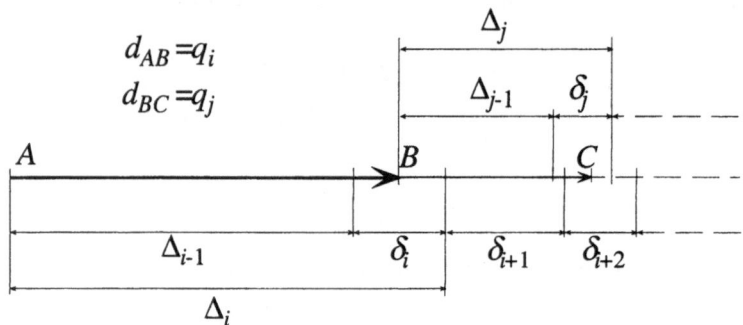

Fig. 2. Composition of distances for same orientation.

To that end, we introduce Algorithm 1 (see also Figure 2), which takes the two distances $d_{AB} = q_i$ and $d_{BC} = q_j$, as well as the structure relations among intervals r_Δ as input.[1] The algorithm first applies the absorption rule to check whether the interval Δ_j can be disregarded with respect to δ_i; if so, $LB = UB = q_i$. Otherwise, to find the upper bound, Δ_j is compared to δ_{i+1} in order to see if Δ_j's outer limit falls within δ_{i+1}. Via recursive calls, the test is repeated for the sum of all successors of δ_i until Δ_j becomes smaller than this sum. The algorithm terminates since, at most, such a sum will eventually include the infinitely big δ_n. The lower bound is computed by comparing Δ_{j-1} with δ_i. If Δ_{j-1} is bigger, then the lower bound must necessarily overcome q_i. The check is repeated recursively with the sum of δ_i and all its successors until Δ_{j-1} becomes smaller.

Algorithm 1
Algorithm for computing the composition of distance relations (same orientation)

begin
   $Input(q_i, q_j, r_\Delta)$;

---

[1] The knowledge given by the structure relations r_Δ supports the evaluation of Boolean predicates. In the case of Algorithm 1, r_Δ gives the results of $\Delta_j \ll \delta_i$, $\Delta_j < \Delta_{\text{inc}}$, and $\Delta_{j-1} < \Delta_{\text{inc}}$.

if $\Delta_j \ll \delta_i$ **then** $UB \leftarrow q_i$; $LB \leftarrow q_i$
      else if $q_i = q_n$ **then** $UB \leftarrow q_i$
            else $FindUB(\delta_{i+1}, i+1, UB)$;
                  $FindLB(\delta_i, i, LB)$ **fi fi**;

$\theta_{AC} \leftarrow \theta_{AB}$;
$Output(LB, UB, \theta_{AC})$

where

proc $FindUB(\Delta_{\mathrm{inc}}, k, \textbf{var } UB) \equiv$
   if $\Delta_j \leq \Delta_{\mathrm{inc}}$ **then** $UB \leftarrow q_k$
         else $FindUB(\Delta_{\mathrm{inc}} + \delta_{k+1}, k+1, UB)$ **fi.**

proc $FindLB(\Delta_{\mathrm{inc}}, k, \textbf{var } LB) \equiv$
   if $\Delta_{j-1} < \Delta_{\mathrm{inc}}$ **then** $LB \leftarrow q_k$
         else $FindLB(\Delta_{\mathrm{inc}} + \delta_{k+1}, k+1, LB)$ **fi.**

end

Opposite Orientation. Now, let us consider the composition of distances in the case of opposite orientation, that is, $\theta_{BC} = \mathrm{opp}(\theta_{AB})$. Algorithm 2 finds upper and lower bounds for the result of composition. It handles separately the three cases $q_i > q_j$, $q_i < q_j$, and $q_i = q_j$, which correspond to three different resulting orientations: equal to θ_{AB}, θ_{BC}, or the logical disjunction of them, respectively. Below, we illustrate how the algorithm manages those three cases.

$q_i > q_j$ Algorithm 2 first applies the absorption rule to check whether the interval Δ_j can be disregarded with respect to δ_i; if so, $LB = UB = q_i$. Otherwise, to find the lower bound (procedure $FindLB$), Δ_j is initially compared to δ_{i-1} to see whether $LB = q_{i-1}$; if not, the test is repeated, via recursive calls, for the sum of all predecessors of δ_{i-1} until Δ_j becomes smaller. The procedure $FindUB$ finds the upper bound exactly in the same way with the only difference that the initial test compares Δ_{j-1} to δ_i (see Figure 3).

$q_i < q_j$ Algorithm 2 first applies the absorption rule to check whether the interval $\Delta_{i..j}$ can be disregarded with respect to δ_0; if so, $LB = UB = q_0$. Otherwise, to find the lower bound (procedure $FindLBopp$), $\Delta_{i+1..j-1}$ is initially compared to δ_0 to see if $LB = q_0$; if not, the test is repeated, via recursive calls, for the sum of all successors of δ_0 until $\Delta_{i+1..j-1}$ becomes smaller. The strategy implemented by the procedure $FindLBopp$ can be informally explained as follows. Since the first distance is smaller than the second one, the first piece of the distance d_{BC} reaches the object A and the remaining part goes beyond A: to calculate how much C overcomes A, the sum of δ_0 and its successors is compared to $\Delta_{i+1..j-1}$. The procedure

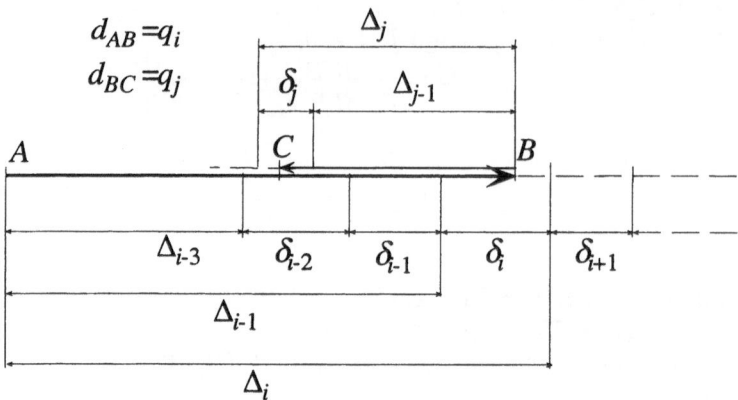

Fig. 3. Composition of distances for opposite orientation ($q_i > q_j$).

FindUBopp works exactly as *FindLBopp* with the only difference that the initial test compares $\Delta_{i..j}$ to δ_1 (see Figure 4).

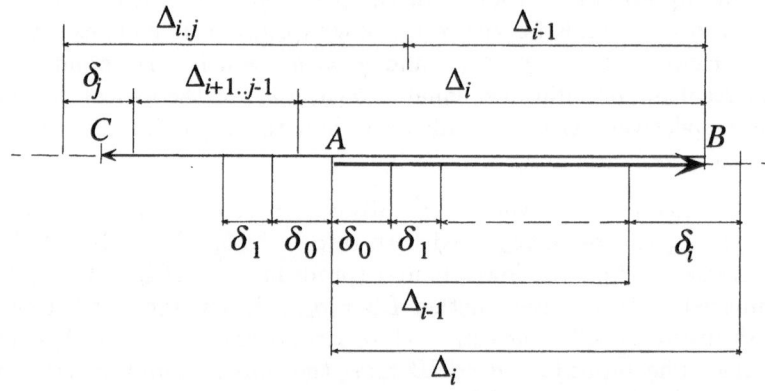

Fig. 4. Composition of distances for opposite orientation ($q_i < q_j$).

$q_i = q_j$ Algorithm 2 first sets $LB = q_0$, hence it proceeds to compute the upper bound (procedure *FindUBeq*). δ_j is initially compared to δ_0 to see if $UB = q_0$; if not, the test is repeated, via recursive calls, for the sum of δ_0 to its successors.

Algorithm 2
Algorithm for computing the composition of distance relations (opposite orientation)

begin
  $Input(q_i, q_j, r_\Delta)$;
  case
    $\Delta_i > \Delta_j$: $\theta_{AC} = \theta_{AB}$;
                if $\Delta_j \ll \delta_i$ **then** $LB \leftarrow q_i$; $UB \leftarrow q_i$
                            else $FindLB(\delta_{i-1}, i-1, LB)$;
                                 $FindUB(\delta_i, i, UB)$ **fi**;
    $\Delta_i < \Delta_j$: $\theta_{AC} = \theta_{BC}$;
                if $\Delta_{i..j} \ll \delta_0$ **then** $LB \leftarrow q_0$; $UB \leftarrow q_0$
                            else $FindLBopp(\delta_0, 0, LB)$;
                                 $FindUBopp(\delta_1, 1, UB)$ **fi**;
    $\Delta_i \cong \Delta_j$: $\theta_{AC} = \theta_{AB} \vee \theta_{BC}$;
                $LB \leftarrow q_0$;
                $FindUBeq(\delta_0, 0, UB)$;
  endcase;
  $Output(LB, UB, \theta_{AC})$

where

proc $FindLB(\Delta_{\text{inc}}, k, \textbf{var } LB) \equiv$
  if $\Delta_j \leq \Delta_{\text{inc}}$ **then** $LB \leftarrow q_k$
              else $FindLB(\Delta_{\text{inc}} + \delta_{k-1}, k-1, LB)$ **fi**.

proc $FindUB(\Delta_{\text{inc}}, k, \textbf{var } UB) \equiv$
  if $\Delta_{j-1} \leq \Delta_{\text{inc}}$ **then** $UB \leftarrow q_k$
              else $FindUB(\Delta_{\text{inc}} + \delta_{k-1}, k-1, UB)$ **fi**.

proc $FindLBopp(\Delta_{\text{inc}}, k, \textbf{var } LB) \equiv$
  if $\Delta_{i+1..j-1} < \Delta_{\text{inc}}$ **then** $LB \leftarrow q_k$
              else $FindLBopp(\Delta_{\text{inc}} + \delta_{k+1}, k+1, LB)$ **fi**.

proc $FindUBopp(\Delta_{\text{inc}}, k, \textbf{var } UB) \equiv$
  if $\Delta_{i..j} < \Delta_{\text{inc}}$ **then** $UB \leftarrow q_k$
              else $FindUBopp(\Delta_{\text{inc}} + \delta_{k+1}, k+1, UB)$ **fi**.

proc $FindUBeq(\Delta_{\text{inc}}, k, \textbf{var } UB) \equiv$
  if $\delta_j < \Delta_{\text{inc}}$ **then** $UB \leftarrow q_k$
              else $FindUBeq(\Delta_{\text{inc}} + \delta_{k+1}, k+1, UB)$ **fi**.

end

Orthogonal Orientation. Let us consider the composition of distances in the case of orthogonal orientation, that is, $\theta_{BC} \in \text{orth}(\theta_{AB})$. In this case, the upper bound tends to be smaller than in the case of same orientation and the lower bound is always the biggest of the distances q_i and q_j. Algorithm 3, which is a modified version of Algorithm 1, computes the upper and lower bounds of distance composition for orthogonal orientation.

To find the upper bound, the procedure $FindUB$ takes into account the following three cases: $\Delta_i \ll \Delta_{\text{inc}}$, $\Delta_i \gg \Delta_{\text{inc}}$, and Δ_i "is comparable to" Δ_{inc}. These three cases are depicted in Figure 5, where we also introduce an auxiliary point H corresponding to the intersection of the segment BC (or its imaginary prolongation) with the radial distance Δ_{i+1} from A. The segment BH can thus be used to decide under which circumstances the upper bound might exceed the current range (in which case a further recursion step will be necessary). To see how this works consider the first step of recursion where $\Delta_{\text{inc}} = \delta_{i+1}$.

If $\Delta_i \ll \delta_{i+1}$ (Figure 5.a), then the segment BH is "slightly" bigger than δ_{i+1}, due to elementary geometric considerations. Therefore, only a value for Δ_j greater than δ_{i+1} can (but not necessarily) increment the resulting distance (next recursive call). Otherwise the upper bound is q_{i+1}.

If $\Delta_i \gg \delta_{i+1}$ (Figure 5.b), then the segment BH is much bigger than δ_{i+1}; hence, it is unlikely that Δ_j is able to increment the resulting distance unless Δ_j is much greater than δ_{i+1}.

In the intermediate cases, that is, when Δ_i is comparable to δ_{i+1} (Figure 5.c), the segment BH is bigger than δ_{i+1}. Hence, Δ_j needs to be considerably bigger than δ_{i+1} in order to overcome the boundary of the next distance range, otherwise the upper bound is q_{i+1}.

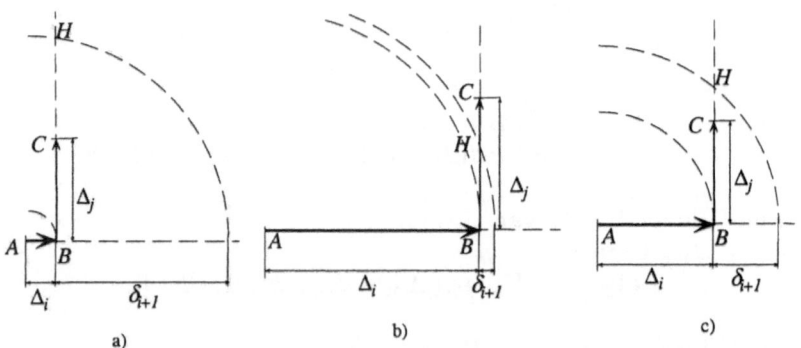

Fig. 5. Composition of distances for orthogonal orientation.

Algorithm 3
Algorithm for computing the composition of distance relations (orthogonal orientation)

<u>begin</u>

$Input(q_i, q_j, r_\Delta)$;
if $\Delta_j \ll \delta_i$ **then** $UB \leftarrow q_i$
           else $FindUB(\delta_{i+1}, i+1, UB)$ **fi**;
if $\Delta_i \geq \Delta_j$ **then** $LB \leftarrow q_i$
           else $LB \leftarrow q_j$ **fi**;
case
    $\Delta_j \gg \Delta_i : \theta_{AC} \leftarrow \theta_{BC}$;
    $\Delta_j \ll \Delta_i : \theta_{AC} \leftarrow \theta_{AB}$;
    else $\theta_{AB} < \theta_{AC} < \theta_{BC}$
endcase;
$Output(LB, UB, \theta_{AC})$

<u>where</u>

<u>proc</u> $FindUB(\Delta_{\text{inc}}, k, $ **<u>var</u>** $UB) \equiv$
    case
        $\Delta_i \ll \Delta_{\text{inc}} : $ **if** $\Delta_j \leq \Delta_{\text{inc}}$
                 then $UB \leftarrow q_k$
                 else $FindUB(\Delta_{\text{inc}} + \delta_{k+1}, k+1, UB)$ **fi**;
        $\Delta_i \gg \Delta_{\text{inc}} : $ **if** $\neg(\Delta_j \gg \Delta_{\text{inc}})$
                 then $UB \leftarrow q_k$
                 else $FindUB(\Delta_{\text{inc}} + \delta_{k+1}, k+1, UB)$ **fi**;
      else if $(\Delta_j < \Delta_{\text{inc}}) \vee (\Delta_j \approx \Delta_{\text{inc}})$
             then $UB \leftarrow q_k$
             else $FindUB(\Delta_{\text{inc}} + \delta_{k+1}, k+1, UB)$ **fi**
    endcase.

  <u>end</u>

Composition for Generic Orientation. As long as the granularity level of orientation relations makes only four distinctions, the application of the algorithms just described is straightforward. If more distinctions are made, however, there are intermediate orientation relations between *same*, *orthogonal*, and *opposite* orientation, for which we have to decide which algorithm gives the best approximation.

For that purpose, it is useful to consider three subranges for intermediate values of θ_{BC}, referring only to the upper half-plane, since all considerations are symmetrical on the axis defined by AB:

1. $\theta_{AB} < \theta_{BC} < \alpha^\perp$;
2. $\alpha^\perp < \theta_{BC} \leq \alpha^*$;
3. $\alpha^* < \theta_{BC} < \text{opp}(\theta_{AB})$,

where $\alpha^{\perp} \in \mathrm{orth}(\theta_{AB})$ and α^* is a particular orientation relation such that the angle between the two lines AB and BC is approximately $120°$.[2] Depending on the number of orientation distinctions, α^* is chosen such that most of its cone lies before $120°$. For example, for eight distinctions α^* is α_2, and for sixteen distinctions α^* is α_5.

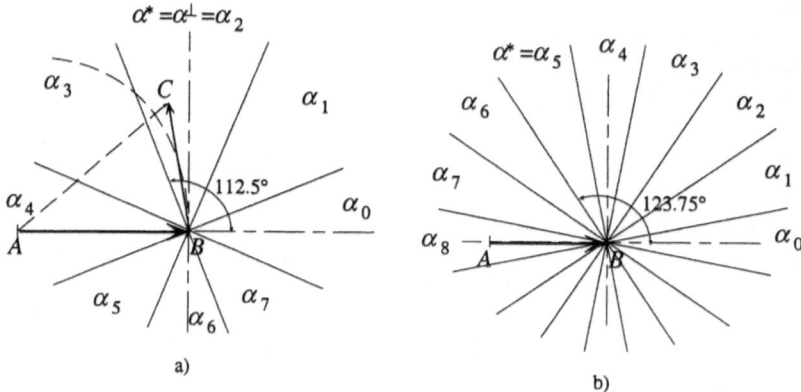

a)   b)

Fig. 6. Composition for intermediate orientation. For eight orientation distinctions (a), a combination of Algorithms 1 and 3 is used for $\theta_{BC} = \alpha_1$ and Algorithm 2 is used for $\theta_{BC} = \alpha_3$. For sixteen orientation distinctions (b), the combination of Algorithms 1 and 3 is used for orientations α_1 through α_3, Algorithm 3 is used for α_5, and Algorithm 2 is used for α_6 and α_7.

For the first subrange, both Algorithms 1 and 3 could be used. Lower and upper bounds inferred by Algorithm 1 are greater than those given by Algorithm 3. The most constrained answer to the composition can be obtained by taking LB from Algorithm 1 and UB from Algorithm 3. For $\theta_{BC} \leq \alpha^*$, the resulting distance d_{AC} is always greater than the biggest distance between d_{AB} and d_{BC}, while, for $\theta_{BC} > \alpha^*$, the resulting distance d_{AC} is always less than the biggest distance between d_{AB} and d_{BC}. Therefore, Algorithm 3 is the most appropriate for the second subrange and Algorithm 2 for the third subrange. The choice of the algorithms for the intermediate orientation relations is illustrated in Figure 6 for the case of eight and sixteen orientation relations.

---

[2] The value $120°$ corresponds to the angle for which the three distances d_{AB}, d_{BC}, and d_{AC} are equal.

3   Path Composition

Now, let us consider the problem of iterating the composition of two positional relations in order to perform more complex spatial reasoning tasks.

A *path* is a sequence of positional relations

$$((d_{AB}, \theta_{AB}), (d_{BC}, \theta_{BC}), (d_{CD}, \theta_{CD}), ...),$$

such that the PO of the i-th relation acts as the RO for the $(i+1)$-th relation.

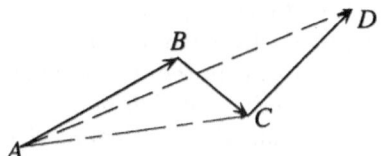

Fig. 7. Composition of positional relations in paths.

With regard to non-trivial paths longer than two positional relations, let us consider the example in Figure 7. The first composition gives a certain range of distances, e.g., $d_{AC} = \{q_1, q_2\}$, and the latter have to be composed with the third relation, e.g., $d_{CD} = q_2$. The distance d_{AD} is calculated by taking into account both possible results for distance d_{AC}: therefore, algorithms for computing $q_1 \oplus q_2$ and $q_2 \oplus q_2$ are run, obtaining two sets of values, e.g., $\{q_2, q_3\}$ and $\{q_3, q_4\}$.

Since any distance relation belonging to the latter sets might be the actual distance d_{AD}, we can conclude that such a distance may assume a value in the *union* of the sets, that is, $d_{AD} = \{q_2, q_3, q_4\}$. Taking the union of the sets can assure us that we are not leaving out any possible result, but this of course has the disadvantage that for long paths the indetermination of the result keeps growing.

A technique that can reduce the indeterminacy of the result is that of "parallel" paths, that is, paths that have common starting and ending points (see example in Figure 8).

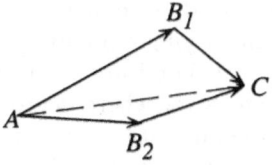

Fig. 8. Parallel paths.

Of course, the positional relations computed along such paths may give different ranges of results, due to the indeterminacy introduced by the algorithms for composition. To avoid conflicting results, we must take into considerations only those values that come from the computations in all parallel paths. If for example the result of the composition of distances in the path AB_1C is in the range $\{q_1, q_2\}$ and in the path AB_2C is in the range $\{q_2, q_3\}$, then the only valid result for distance composition is $d_{AC} = q_2$. As a general rule, parallel paths must be checked for consistency by taking the *intersection* of the sets of values obtained along different paths.

The technique of parallel paths can be used to reduce indetermination in path composition. In fact, distance d_{AD}, which in Figure 7 was computed as $d_{AC} \oplus d_{CD}$, can be computed alternatively by the composition $d_{AB} \oplus d_{BD}$ (see Figure 9). The results for d_{AD} in the two parallel paths ACD and ABD can be combined by taking the intersection of values. Checking consistency among various ways of obtaining the final composition along a given path means that we perform the intersection of several results, reducing at the same time indetermination.

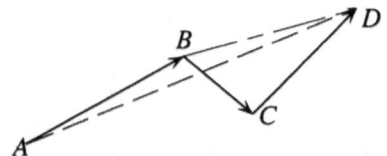

Fig. 9. Alternative way of composition in paths.

4   Integrating Qualitative and Fuzzy Knowledge

The propagation of qualitative reasoning with positional information leads inevitably to a deterioration of the result since the range of possible qualitative answers tends to increase. Hereafter, we investigate how to reduce the uncertainty of the result if qualitative information is integrated with fuzzy knowledge about *degrees of likelihood* [19] of each qualitative distance value. Such fuzzy knowledge is additional knowledge that may come from different sources: it could be user-defined, determined by context, or inferred from the geometric structure of the underlying domain. In this section, we will generally assume that degrees of likelihood are given, while in Section 5, we will see an application of the framework where degrees of likelihood are inferred from the structure of a particular distance system.

Let us define a *fuzzy qualitative positional relation* between two points A and B as

$$(D_{AB}, \Theta_{AB}),$$

where D_{AB} and Θ_{AB} are vectors containing the distributions of degrees of likelihood for all qualitative distance and orientation symbols.

More in detail, the distance relation D_{AB} is defined as:

$$D_{AB} = \begin{pmatrix} q_0 : \gamma_0 \\ q_1 : \gamma_1 \\ \cdot \\ \cdot \\ \cdot \\ q_n : \gamma_n \end{pmatrix},$$

where each γ_i gives the degree of likelihood of the distance symbol q_i. The structure of Θ_{AB} is similar to that of D_{AB}, containing degrees of likelihood of orientation symbols. In the rest of this paper, we concentrate on distance relations only.

For example, if

$$D_{AB} = \begin{pmatrix} q_0 : 0 \\ q_1 : 0.8 \\ q_2 : 0.5 \\ q_3 : 0 \\ q_4 : 0 \end{pmatrix},$$

then the distance between points A and B is q_1 with a degree of likelihood 0.8 and q_2 with a degree of likelihood 0.5. Other distance symbols (i.e., q_0, q_3, q_4) have a degree of likelihood equal to zero.

The composition of two distance relations D_{AB} and D_{BC} is carried out by applying the algorithms for composition (Section 2.2) to each possible combination of distance symbols and by aggregating the results with a strategy that takes into account degrees of likelihood with which each distance symbol participates in the relation. A general strategy is described in the following.

Given

$$D_{AB} = \begin{pmatrix} q_0 : \lambda_0 \\ \cdot \\ \cdot \\ \cdot \\ q_n : \lambda_n \end{pmatrix},$$

and

$$D_{BC} = \begin{pmatrix} q_0 : \mu_0 \\ \cdot \\ \cdot \\ \cdot \\ q_n : \mu_n \end{pmatrix},$$

the result of the composition $D_{AB} \oplus D_{BC}$ is indicated by:

$$D_{AC} = \begin{pmatrix} q_0 : \nu_0 \\ \cdot \\ \cdot \\ \cdot \\ q_n : \nu_n \end{pmatrix}.$$

In order to find D_{AC}, we have to be able to calculate the basic composition

$$(q_i : \lambda_i) \oplus (q_j : \mu_j),$$

for all i and j, totaling $(n + 1)^2$ compositions. Each of them gives a partial result for D_{AC} equal to:

$$\begin{pmatrix} q_0 : \nu_0^{ij} \\ \cdot \\ \cdot \\ q_n : \nu_n^{ij} \end{pmatrix}.$$

The above basic composition is carried out by (**Step 1**) calculating

$$q_i \oplus q_j = \{q_x, ..., q_y\}, \text{with } y \geq x,$$

where $q_x...q_y$ is the range of possible values for the composition of q_i and q_j (which is given by the algorithms of Section 2.2 substituting $LB = q_x$ and $UB = q_y$).

Then (**Step 2**), the degrees of likelihood to be assigned to the above results are calculated using the fuzzy intersection[3] $\lambda_i \cap \mu_j$. Then we distribute the fuzzy intersection over symbols $q_x...q_y$ by using a *distribution function* ϕ such that:

$$\phi(\lambda_i \cap \mu_j) = \{\nu_x^{ij}, ..., \nu_y^{ij}\}.$$

The distribution function ϕ can be of different forms, depending on the application domain: for example, it can divide the fuzzy intersection by the cardinality of the resulting set of qualitative distances or can simply assign the same value of the fuzzy intersection to each qualitative distance. A particular kind of ϕ function will be used in Section 5.

For example, if as an instance of fuzzy union we take the minimum, and as an instance of ϕ function we take a division by the cardinality, we obtain:

$$\nu_x^{ij} = ... = \nu_y^{ij} = \min(\lambda_i, \mu_j)/(y - x + 1).$$

Once every aforementioned basic composition is performed, we have to aggregate all the contributions for all i and j. Each degree of likelihood ν_k

---

[3] Various formulas have been proposed in the literature [19] for fuzzy intersection. Given the degrees of likelihood λ and μ, valid formulas for $\lambda \cap \mu$ are:

(a) (minimum) $\min(\lambda, \mu)$;

(b) (Yager) $\begin{cases} \lambda + \mu - 1 & \text{if } \lambda + \mu > 1, \\ 0 & \text{otherwise}; \end{cases}$

(c) (1st Hamacher) $\lambda \cdot \mu$;

(d) (2nd Hamacher) $(\lambda \cdot \mu)/(2 - \lambda - \mu + \lambda \cdot \mu)$.

contained in the result of composition D_{AC} is calculated by taking the fuzzy union[4] of all ν_k^{ij}, that is:

$$D_{AC} = \begin{pmatrix} q_0 : \bigcup_{ij} \nu_0^{ij} \\ \cdot \\ \cdot \\ \cdot \\ q_n : \bigcup_{ij} \nu_n^{ij} \end{pmatrix}.$$

The simplest form of fuzzy union is taking the maximum of values. Functions ϕ and fuzzy union and intersection add flexibility to the framework proposed in this paper, providing different ways of combining degrees of likelihood and distributing them over various distance values. The form that these functions assume can be adjusted empirically to fit the particular sources of knowledge.

5   Fuzzy Knowledge Coming from Particular Distance Systems

In this section, we apply the general framework of previous section to the case where information about the ratio among various intervals of the underlying distance system is available.

Let us refer to a distance system as the one illustrated in Figure 10. In such a system the structure relations among intervals are known in terms of the ratios among the lengths of intervals, e.g., $\|\delta_0\|/\|\delta_1\| = 5/6$, $\|\delta_1\|/\|\delta_2\| = 6/7$, $\|\delta_2\|/\|\delta_3\| = 7/8$, $\|\delta_3\|/\|\delta_4\| = 8/9$, $\|\delta_5\| = +\infty$.

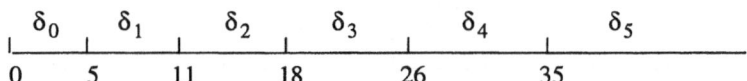

Fig. 10. A distance system where the ratio among intervals is known.

Based on this distance system, we carry out the composition of three consecutive distances with values $d_{AB} = q_1$, $d_{BC} = q_2$, $d_{CD} = q_1$. In terms

---

[4] Analogously to fuzzy intersection, given the degrees of likelihood λ and μ, valid formulas for fuzzy union $\lambda \cup \mu$ are:

(a) (maximum) $\max(\lambda, \mu)$;

(b) (Yager) $\begin{cases} \lambda + \mu & \text{if } \lambda + \mu < 1, \\ 1 & \text{otherwise}; \end{cases}$

(c) (1st Hamacher) $\lambda + \mu - \lambda \cdot \mu$;

(d) (2nd Hamacher) $(\lambda + \mu)/(1 + \lambda \cdot \mu)$.

of the formalism of previous section, we first perform the fuzzy composition between

$$D_{AB} = \begin{pmatrix} q_0 : 0 \\ q_1 : 1 \\ q_2 : 0 \\ q_3 : 0 \\ q_4 : 0 \\ q_5 : 0 \end{pmatrix},$$

and

$$D_{BC} = \begin{pmatrix} q_0 : 0 \\ q_1 : 0 \\ q_2 : 1 \\ q_3 : 0 \\ q_4 : 0 \\ q_5 : 0 \end{pmatrix}.$$

In Step 1, we apply the algorithm for the composition of distance relations in case of same orientation (Algorithm 1)[5]. For this particular distance system,

$$q_1 \oplus q_2 = \{q_2, q_3, q_4\}.$$

The above result is graphically explained in Figure 11. In Step 2 of the method, to perform the fuzzy intersection, we consider the minimum of the degrees of likelihood: in this case, it is trivially $min(1, 1) = 1$.

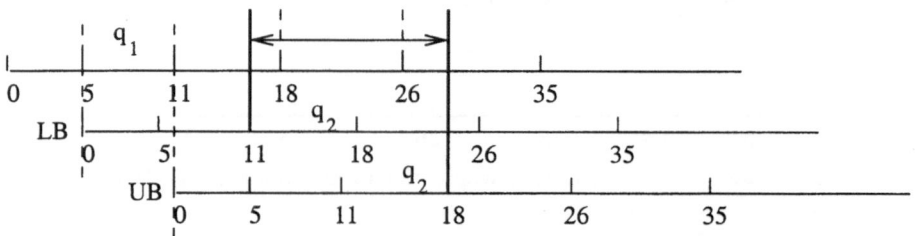

Fig. 11. Graphical explanation of $q_1 \oplus q_2 = \{q_2, q_3, q_4\}$.

The figure also suggests that the result can be refined by assigning to each distance symbol of the result a degree of likelihood that is proportional to the length of the corresponding interval. The point C may actually fall in an interval that overlaps δ_2 by a length of two units, completely contains δ_3 (which is 8 units long), and overlaps δ_4 by a length of three units. Therefore, this suggests to use as the ϕ function a multiplication by 2/13, 8/13, 3/13, which express the ratios between the part of interval related to a certain

---

[5] If orientation is different, other algorithms can be applied without restrictions.

distance symbol and the whole resulting interval. Since there are no other contributions, in this case, we simply have:

$$
D_{AC} = \begin{pmatrix} q_0 : 0 \\ q_1 : 0 \\ q_2 : 1 \cdot 2/13 \\ q_3 : 1 \cdot 8/13 \\ q_4 : 1 \cdot 3/13 \\ q_5 : 0 \end{pmatrix} = \begin{pmatrix} q_0 : 0 \\ q_1 : 0 \\ q_2 : 0.154 \\ q_3 : 0.615 \\ q_4 : 0.231 \\ q_5 : 0 \end{pmatrix} .
$$

Now, we iterate the composition of D_{AC} with

$$
D_{CD} = \begin{pmatrix} q_0 : 0 \\ q_1 : 1 \\ q_2 : 0 \\ q_3 : 0 \\ q_4 : 0 \\ q_5 : 0 \end{pmatrix} .
$$

In Step 1, the application of Algorithm 1 gives:

$$
q_2 \oplus q_1 = \{q_2, q_3, q_4\},
$$

$$
q_3 \oplus q_1 = \{q_3, q_4, q_5\},
$$

$$
q_4 \oplus q_1 = \{q_4, q_5\}.
$$

The composition $q_2 \oplus q_1$ was already illustrated in Figure 11, while compositions $q_3 \oplus q_1$ and $q_4 \oplus q_1$ are illustrated in Figures 12 and 13, respectively (note that commutativity holds for composition).

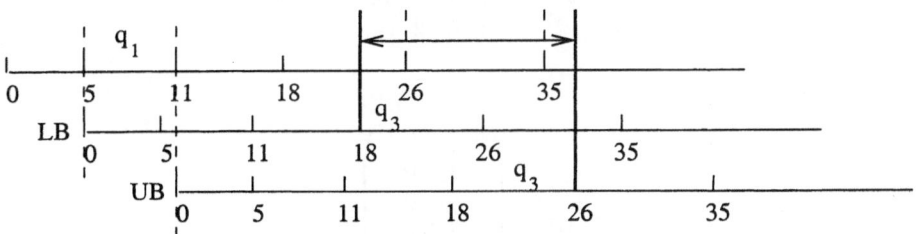

Fig. 12. Graphical explanation of $q_1 \oplus q_3 = \{q_3, q_4, q_5\}$.

Step 2 of the method consists of the application of the minimum function: $\min(\lambda_2, \mu_1) = \min(0.154, 1) = 0.154$, $\min(\lambda_3, \mu_1) = \min(0.615, 1) = 0.615$, and $\min(\lambda_4, \mu_1) = \min(0.231, 1) = 0.231$.

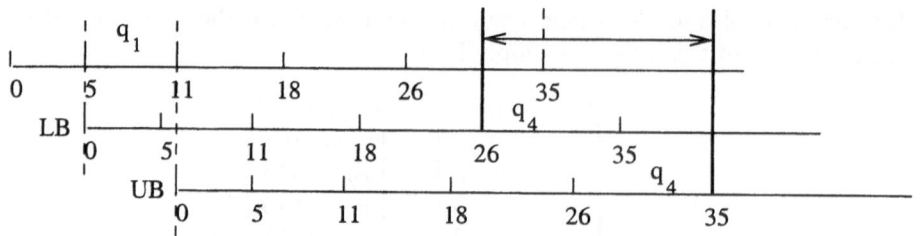

Fig. 13. Graphical explanation of $q_1 \oplus q_4 = \{q_4, q_5\}$.

By using the already discussed criteria for the distribution function, we obtain:

$$\nu_2^{21} = 0.154 \cdot 2/13$$
$$\nu_3^{21} = 0.154 \cdot 8/13$$
$$\nu_4^{21} = 0.154 \cdot 3/13.$$

Therefore, the contribution of the first composition to the result is:

$$D'_{AD} = \begin{pmatrix} q_0 : 0 \\ q_1 : 0 \\ q_2 : 0.024 \\ q_3 : 0.095 \\ q_4 : 0.036 \\ q_5 : 0 \end{pmatrix}.$$

For the second contribution, we obtain:

$$\nu_3^{31} = 0.625 \cdot 3/14$$
$$\nu_4^{31} = 0.625 \cdot 9/14$$
$$\nu_5^{31} = 0.625 \cdot 2/14,$$

and, therefore:

$$D''_{AD} = \begin{pmatrix} q_0 : 0 \\ q_1 : 0 \\ q_2 : 0 \\ q_3 : 0.132 \\ q_4 : 0.395 \\ q_5 : 0.098 \end{pmatrix}.$$

For the third contribution, we have:

$$\nu_4^{41} = 0.231 \cdot 4/15$$
$$\nu_5^{41} = 0.231 \cdot 11/15,$$

and, therefore:

$$D'''_{AD} = \begin{pmatrix} q_0 : 0 \\ q_1 : 0 \\ q_2 : 0 \\ q_3 : 0 \\ q_4 : 0.062 \\ q_5 : 0.169 \end{pmatrix}.$$

For integrating the three contributions above, we make use of the Yager formula for fuzzy union, that is, we add the degrees of likelihood, obtaining:

$$D_{AD} = \begin{pmatrix} q_0 : 0 \\ q_1 : 0 \\ q_2 : 0.024 \\ q_3 : 0.227 \\ q_4 : 0.493 \\ q_5 : 0.257 \end{pmatrix}.$$

In the method illustrated in this section, the sum of degrees of likelihood in each vector is 1. By using the knowledge about ratios between lengths of intervals, we are able to assign different degrees of likelihood to the result of iterated composition. In the example above, we can conclude that distance d_{AD} is more likely to be q_4 (with a degree 0.493), while the value q_2 is rather unlikely to be verified (with a degree 0.024).

6   Conclusions

Qualitative spatial reasoning about positional relations is not much used in practice since the iteration of basic steps of composition along a path rapidly enlarges the indetermination of the result.

In this chapter, we face the above problem by coupling qualitative with fuzzy knowledge. Fuzzy knowledge may come from different sources: as an example, we considered fuzzy knowledge coming from the mapping from qualitative distance symbols to geometric intervals. We have shown that in this way we obtain indications about which is the qualitative distance that is more likely to be the result of composition among the various possible purely qualitative answers.

The proposed framework has several degrees of flexibility to accommodate various kinds of fuzzy knowledge, namely, by choosing a formula to calculate fuzzy union and intersection of degrees of likelihood, and by choosing a distribution function to give different weights to qualitative values. As an example, we considered a distance system where it was convenient to take the minimum as the fuzzy intersection, the Yager formula as the fuzzy union, and a proportion to the lengths of intervals as a distribution function.

Further research will be about the criteria to combine degrees of likelihood in parallel paths with fuzzy knowledge. Also, fuzzy knowledge can be associated to orientation relations and integrated to the present work.

Acknowledgments

This work has been supported by the Italian MURST project "Rappresentazione ed elaborazione di dati spaziali nei Sistemi Informativi Territoriali".

References

1. James F. Allen. Maintaining knowledge about temporal intervals. *Communications of the ACM*, 26(11):832–843, November 1983.
2. David Altman. Fuzzy set theoretic approaches for handling imprecision in spatial analysis. *International Journal of Geographical Information Systems*, 8(3):271–289, 1994.
3. Eliseo Clementini and Paolino Di Felice. Spatial operators. *ACM SIGMOD Record*, 29(3):31–38, 2000.
4. Eliseo Clementini, Paolino Di Felice, and Daniel Hernández. Qualitative representation of positional information. *Artificial Intelligence*, 95:317–356, 1997.
5. A. G. Cohn. The challenge of qualitative spatial reasoning. *Computing Surveys*, 27(3):323–327, 1995.
6. A. G. Cohn and S. M. Hazarika. Qualitative spatial representation and reasoning: An overview. *Fundamenta Informaticae*, 43:2–32, 2001.
7. R. M. Downs and D. Stea. Cognitive maps and spatial behavior: Process and products. In *Image and Environment: Cognitive Mapping and Spatial Behavior* [8], pages 8–26.
8. R. M. Downs and D. Stea, editors. *Image and Environment: Cognitive Mapping and Spatial Behavior*. Aldine, Chicago, 1973.
9. Soumitra Dutta. Approximate spatial reasoning: Integrating qualitative and quantitative constraints. *International Journal of Approximate Reasoning*, 5:307–331, 1991.
10. Kenneth D. Forbus, Paul Nielsen, and Boi Faltings. Qualitative spatial reasoning: The clock project. *Artificial Intelligence*, 51:417–471, 1991.
11. H.W. Guesgen and J. Albrecht. Imprecise reasoning in geographic information systems. *Fuzzy Sets and Systems (Special Issue on Uncertainty Management in Spatial Data and GIS)*, 113(1):121–131, 2000.
12. R. A. Hart and G. T. Moore. The development of spatial cognition: A review. In Downs and Stea [8].
13. Jerry R. Hobbs. Granularity. In Aravind Joshi, editor, *Proceedings of the Ninth International Joint Conference on Artificial Intelligence*, pages 432–435, San Mateo, CA, 1985. International Joint Conferences on Artificial Intelligence, Inc., Morgan Kaufmann.
14. A. Isli and R. Moratz. Qualitative spatial representation and reasoning: Algebraic models for relative position. Technical report, Fachbereich Informatik, Universitaet Hamburg, 1999.
15. Joaquim A. Jorge and Dragos Vaida. A fuzzy relational path algebra for distances and directions. In *Proceedings of the ECAI-96 Workshop on Representation and Processing of Spatial Expressions*, 1996.
16. Jiming Liu. A method of spatial reasoning based on qualitative trigonometry. *Artificial Intelligence*, 98:137–168, 1998.
17. K. Lynch. *The Image of the City*. The MIT Press, Cambridge, MA, 1960.

18. M. L. Mavrovouniotis and G. Stephanopoulos. Formal order-of-magnitude reasoning in process engineering. *Computer Chemical Engineering*, 12:867–880, 1988.
19. H.-J. Zimmermann. *Fuzzy set theory and its applications*. Kluwer, Dordrecht, 1992.
20. Kai Zimmermann. Enhancing qualitative spatial reasoning—combining orientation and distance. In Andrew U. Frank and Irene Campari, editors, *Spatial Information Theory. A Theoretical Basis for GIS. European Conference, COSIT'93*, volume 716 of *Lecture Notes in Computer Science*, pages 69–76, Berlin, September 1993. Springer.
21. Kai Zimmermann and Christian Freksa. Qualitative spatial reasoning using orientation, distance, and path knowledge. *Applied Intelligence*, 6:49–58, 1996.

Spatial Relations
Based on Dominance of Fuzzy Sets

Les Sztandera

Computer Science Department
Philadelphia University
Philadelphia, PA 19144-5497
SztanderaL @philau.edu

Abstract. Spatial relationships between regions in an image play an important role in scene understanding. Humans are able to quickly ascertain the relationship between two objects, for example "B is to the right of A," or "B is in front of A," but this has turned out to be a somewhat illusive task for automation. When the objects in a scene are represented by crisp sets, the all-or-nothing definitions of the subsets actually add to the problem of generating such relational descriptions. It is our belief that definitions of spatial relationships based on fuzzy set theory, coupled with a fuzzy segmentation will yield realistic results. This chapter presents an approach at defining spatial relationships among fuzzy subsets of an image plane. The idea is to project the fuzzy subsets onto two orthogonal coordinate axes and to utilize fuzzy set theoretic dominance relations to capture the approximate relationships. Simulation results are provided to corroborate the theory.

Keywords. Spatial relations, dominance of fuzzy sets, separation measure.

1 Spatial Relations

1.1 Introduction

The problem of specifying the spatial relations between objects in a scene is a subpart of the much more general problem of picture description. Our task logically begins with a two-dimensional encoding of a scene as a picture, which is of course a basic kind of description. We desire to transform this into some "higher level" specification of what it is a picture of, with respect to the limited semantic world of spatial relations. A first step in this process would be to segment the picture into what we call the objects. These objects will hopefully correspond to what would strike an observer as the separable parts of the scene. Objects are complex entities composed of lines, edges and regions in various relations and with various properties, e.g., shape, size and texture. Our picture description will consist of objects or the constellation of relations and properties representing objects - and the relations between them. We will delineate two types of relations between objects:

(i)   those involving comparison of the properties of objects (larger, darker, smoother), and

(ii) those involving their relative position (above, near, to the left of).

Type (ii) could be considered a subset of type (i) if the location of an object were part of its property list, but because our concern is with spatial relations, we will find it convenient to maintain the distinction, and limit ourselves to type (ii) relations.

1.2 Modeling of Spatial Relations

Relations involving relative position of the object are [1]:

1. LEFT OF
2. RIGHT OF
3. BESIDE (alongside, next to)
4. ABOVE (over, higher than, on top of)
5. BELOW (under, underneath, lower than)
6. BEHIND (in back of)
7. IN FRONT OF
8. NEAR (close to)
9. FAR
10. TOUCHING
11. BETWEEN
12. INSIDE (within)
13. OUTSIDE

These should be read as, e.g.:
BELOW(A,B)= "A is below B," BETWEEN (A,B,C)= "A is between B and C."

Above is a list of names for spatial relations. This is intended to be an exhaustive list of "primitive" names; that is, the many relations not listed here that can hold among objects in a scene are describable only as combinations of these, e.g., "to the left and above." A few preliminary observations about the meanings of these terms are made here:

1. To say that one object is behind (in front of) another object in a scene makes sense only if the scene is interpreted as representing a three-dimensional situation. The other eleven relations could simply pertain to the deployment of the parts of the picture in the plane, with no particular reference to the depth of real world.

2. BETWEEN is of necessity a ternary relation; all the other relations are most conveniently described as being binary.

3. Allowing the possibility of objects touching introduces much extra complexity into the segmentation problem mentioned earlier. The segmenter must decide whether a particular part of the picture is one object, or two objects touching. This is usually not a simple problem.

4. We can look at these terms with respect to the standard mathematical properties of relations:

reflexive - TOUCHING;
symmetric - NEAR, FAR, BESIDE;
anti-symmetric and transitive - LEFT, RIGHT, ABOVE, BELOW,
                              BEHIND, IN FRONT, INSIDE, OUTSIDE.

Those relations that are anti-symmetric and transitive are strict partial order relations.

5. Except for BESIDE and TOUCHING, each binary relation is coupled with a relation that is its semantic inverse; for example, ABOVE (A,B), ~ BELOW (B,A). Note that two objects might be positioned so that neither of a relation pair is applicable, and none of the relations describing them would be a total order relation, that is, none is anti-reflexive, anti-symmetric and transitive. Our list could be expended to include terms complementary to BESIDE and TOUCHING in meaning, but according to "The Pan Dictionary of Synonyms and Antonyms" [3] these ideas could only be expressed as negations of the given terms NOT BESIDE, NOT TOUCHING, and so are not primitive enough to be included.

6. There exist mathematical definitions of INSIDE and OUTSIDE as strictly topological properties of objects, and there exist algorithms for determining whether one object in a digital picture is INSIDE (OUTSIDE) another .

7. The all-or-none nature of standard relational mathematics is inadequate to the task of modeling human judgments. Fig. 1 exemplifies this. The LEFT relation is clearly applicable in a) case, and clearly inapplicable in f) case. Defining LEFT as a fuzzy relation would allow the specification of the degree of relatedness for the cases in between. The characteristic function of LEFT would map the a) objects into 1 or some number close to 1, with each successive pair mapped into a smaller number, until f of the case f) would be equal LEFT to or close to 0.

1.3 Comparison of Definitions of Spatial Relations

1.3.1 Winston's Model

A landmark work attempting to quantify spatial relations has been the work of Winston [2]. His primary interest was in creating a program that could learn to recognize line drawing representations of structures by building an abstract description of a given line drawing and examining the applicability of various in-

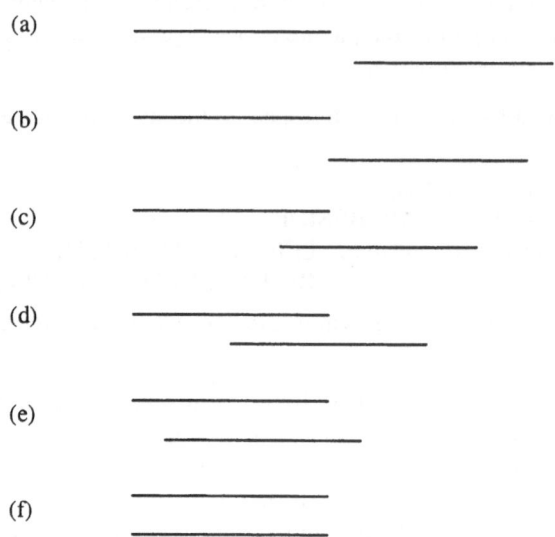

Fig. 1. An example of a relation "to the left of " (based on [2]).

ternalized structure descriptions to it, so as to be able to say, for example, "This is an arch (or table, etc.)." Fundamental to this endeavor is the ability to build useful scene descriptions, and it is this ability, rather than the use of the descriptions for machine learning, that concerns us.

Winston's program accepts as input line drawings of three-dimensional scenes. A program called SEE segments the picture into separate objects by a series of operations on local features such as collinear or parallel line segments and type of vertex present (e.g., L, Y, etc.). These operations link pairs of regions together, and the most strongly linked regions become associated as faces of a single object. This is all done under the assumption that the scene represents "three dimensional solids formed by planes," viewed from above and resting on a plane surface. The relations that Winston uses to describe the scenes thus segmented are ABOVE, SUPPORTS, IN FRONT OF, LEFT, RIGHT, and MARRYS. As a first step in assigning these relationships the program finds a bottom border of each figure, according to the vertical displacement of the vertices of the figure and their type. Once it has good candidates for bottom lines (and for which face of the object is resting on another object or the table), the program builds a network of ABOVE and SUPPORTS relations between the objects according to the rule:

"Some face of A is below bottom line of B ⇔ SUPPORTS(A,B)."

In Winston's world, (SUPPORTS(A,B) ⇔ ABOVE(B,A)), which is reasonable, since his scenes represent a situation in which gravity exists. The IN-FRONT-OF relation is defined as follows: "A partially obscures B ∧ ABOVE(A,B) ⇔ IN-FRONT-OF(A,B)." That A partially obscures B should be part of the network description of the scene that SEE generated. The rest of the expression derives from the simple observation that looking from above, if B is on top of A it will of necessity obscure at least part of A.

The following is a first pass definition of LEFT (XCOR - the object's horizontal position):

> if XCOR (all vertices of A) < XCOR (all vertices of B)
>   then LEFT(A,B);
> else if XCOR (center of area of A) < XCOR (all vertices of B)
>   then LEFT(A,B);
> else FAIL;

Note that the "origin" is in the lower left corner of the scene, that is, values of x increase to the right, and values of y increase upward.

This rule is quite effective in many cases, as can be seen in Fig. 1(a-f). This rule says that the LEFT relation holds only in Fig. 1(a-d). To my intuition, that is as reasonable a cutoff point as any. Let f_A (x,y) be the characteristic function that associates each element of S x S, where S is some object space, with some number in the interval [0, 1]. If A is a binary relation over pairs of objects from the object space S, then the value of f_A closer to 1 indicates a greater relatedness between A the objects. Thus, the characteristic function of LEFT would map the (1a) objects (however one maps objects) into 1 or some number close to 1, with each successive pair mapped into a smaller number, until f_{LEFT} of (1f) would be equal to or close to 0.

We could introduce two thresholds α and β, $0 < \alpha < 1$, $0 < \beta < 1$, $\alpha > \beta$ such that $f_A \leq \beta$ (the relation A does not hold), $f_A \geq \alpha$ (the relation A does hold): $\alpha < f_A < \beta$ (the relatedness is A indeterminate). This approach could be applied to Fig. 1 quite nicely. It is an attractive approach because: i) the response it gives seems more humanlike, and ii) the strictness or laxity of a relational definition is localized in the threshold selection. However, Winston's definition has the interesting property of allowing the situation in Fig. 2(a), where B is to the right of A, but A is not to the left of B (note that this is counter to our observation 5 in section 1.2). In Fig. 2(b) A is to the left, of B, but B is not to the right of A. Winston notes this, and makes an appropriate modification to his rule, but the reader should contemplate Fig. 2 and ask himself whether LEFT and RIGHT are always symmetric.

The MARRYS relation is used to describe "objects that touch each other and have at least one common edge." It is deduced from classifying vertices along common borders of objects (as found by SEE), to decide if the objects abut or overlap.

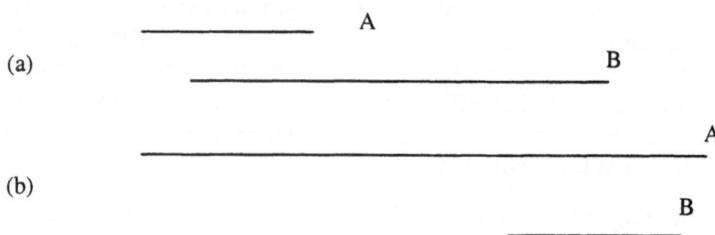

(a)

(b)

Fig. 2. Examples of relations "to the left (right) of " (based on [2]).

Context Sensitivity

It would seem desirable in any complex relational system to have the relations between parts be sensitive to information additional to the property lists of parts in question. This kind of sensitivity is almost mandatory in any model that aspires to psychological relevance. In our particular area of concern, this means that spatial relations between two parts of a scene can be affected by what the rest of the scene looks like. A possibly counterintuitive example (here, it means not congruent to human performance) can be fabricated from the simple version of Winston's rule given above. In both Figs. 3a and 3b, the objects A and B occupy the same relative position, and by some comparison of the Y-coordinates, A would be above B. However, in Winston's scheme, where ABOVE is defined in terms of SUPPORTS, A is ABOVE B in 3a, but not in 3b, since in 3a we have:

SUPPORTS(B,X) ∧ SUPPORTS(X,A) ⇒ SUPPORTS(B,A) ⇒ ABOVE(A,B).

There is no chance of the ABOVE relation applying in 3b, since A and B have no SUPPORTS in common. As an example of the role of the set of the observer in scene description by humans, these conflicting results are quite reasonable. If the pictures are thought of as representing three-dimensional solids resting on an imaginary table and seen from a side-on view, then Winston's verdict is acceptable. But, if one somehow erases that interpretation and instead looks at the objects as geometrical figures disposed on a plane, then A seems to be above B in both scenes, and the interposition of the other geometric figures seems to have

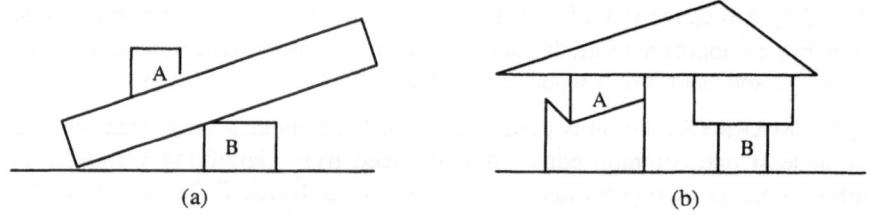

(a)   (b)

Fig. 3. Example of the relation "above."

little effect on the interpretation. An important part of our design will be to thoughtfully define spatial relations in terms of more primitive relations and properties so that they will exhibit this sensitivity to the context of the rest of the scene, and as a more ambitious goal, so that they will be sensitive to different ways of interpreting the scene.

1.3.2 Psychological Models

We will now review some of the linguistic and psychological literature concerning the semantic nature of these terms of spatial relation, with particular attention to a recent attempt to embody this knowledge in a process model of human perceptual encoding.

Clark and Chase -Relational Encoding

"Near/far " are paired members of the class of terms of measure or quality, since they contain a notion of position on some evaluative dimension [4]. As members of this class they have the quality that one of the pair is marked. That is, "farness" describes both: i) the whole dimension of distance (its nominative sense), and ii) the more extended end of that dimension (its contrastive sense) [5]. It is the neutral, or unmarked, term. "Nearness," on the other hand, has only a contrastive sense: it describes only the unexpended end of the spectrum. Thinking again in terms of fuzzy relations, we can ask if the characteristic function for NEAR is related in any simple way to that for FAR, for example, is $\mu(f_{NEAR}) = 1 - \mu(f_{FAR})$ This particular form of relatedness is probably only applicable to NEAR/FAR, since this is the only pair whose numbers are so clearly quantitatively related, as in "The farther away an object is, the less near it is." "Nearness" is marked as "not far." This is illustrated below, by the anomalous nature of (ii):

(i) How far is it from A to B? (ii) How near is it from A to B?

"Above/below," "left/right," "behind/in front of " are prima facie not adjectives of measure, for example: A is nearer than B, but not A is more above (more behind) B. Clark and Chase [6] argue, however, that at least "above/below" still possess the attribute of markedness. They point out that "iii) A is above B and iv) B is below A" are not strictly synonymous, in that the reference point changes from B in iii) to A in iv). They couple this with the observation that "height" is measured upward from a point of reference at the bottom to assert that iii) is the neutral (unmarked) description of A being on top of B, since in iii) the lower object is the reference point. This theorizing is given an impressive support by the experiments they performed. In [6], there is a systematic presentation of these experiments. The authors attack a more general problem of great interest here, namely, that of the relation between human encoding of a simple picture and the encoding of a linguistic description of that picture. They have three major hypotheses about this relationship that they wish to test:

A. According to some a priori conception of perceptual space, people prefer some coding of pictures over others, regardless of their language

B. A descriptive sentence presented to a subject before the picture will normally determine his initial encoding of the picture, even if it is not the preferred encoding of Hypothesis A

C. Sometimes the properties of the picture so strongly suggest some particular encoding that Hypothesis B may be overridden.

This concludes my review of Clark and Chase's approach to the problem of encoding perceptual events. The authors are not particularly concerned with what seems to me to be a major part of our understanding of spatial relations, and certainly of our linguistic description of spatial relations. That is, what about the manifest psychological differences in the encoding of horizontal and vertical relations? Clark's apparent lack of interest may stem from the fact that although left and right are a respectable pair of comparative terms, they are not well behaved like above-below, front-back, etc., in that:

(i)   their stimulus dimension has no apparent zero point, so that
(ii)   one of the terms is not more extended than the other, so that
(iii) the terms do not display the quality of markedness.

To encode the situation of two objects in some horizontal relation, people would probably proceed according to:

Rule 1. People normally code pictures in positive terms;
Rule 2. People refer to the locations of objects positively,

where upward and forward from the observer are positive directions; but if there is no previous context, and if the objects are of equal perceptual prominence, then I would not care even to guess at what the preferred encoding would be.

1.4 Summary

With respect to the problem of programming a model of spatial relations, we have come up with many more questions than answers. We have discussed the nameable relations, and found them to be quite varied in semantic content. It appeared at the outset that some sort of relational data structure would be appropriate to represent these relations, and we showed that a fuzzy relational structure would be particularly appropriate to model the continuum from applicability to nonapplicability that characterizes most of these relations.

The psychological literature shed some light on how to build a model. It told us that a person's perceptual codes are a function of the three factors of his own propensities, how he intends to use the perception, and the prominent qualities of the perceptual event itself. We gained some insight into each step of the process of perceiving:

1. Before we can have relations between objects we must have some representation for the objects themselves. An object in a scene could be a data structure of the relations between the parts of the object. A big step in sophistication requiring a great wealth of knowledge of the world and overhead in structure matching would be to recode these representations into the names of objects. It would be more feasible for our model to be given names for objects, with knowledge of the properties of an object having a given name. Our model would then be applicable only to the relations between certain kinds of geometric figures, namely those we told it about.

2. If we have some representation for the objects in a scene, we then need to decide what relations apply to those objects. One or the other of the relation pairs INSIDE/OUTSIDE and NEAR/FAR are applicable to every pair of objects, with the possible exception of intermediate cases of NEAR/FAR. It might be useful to have characteristic functions that give us indices of applicability for the relation pairs ABOVE/BELOW, LEFT/RIGHT, IN FRONT OF/BEHIND. As a matter for further study, it might turn out that the marked/unmarked nature of some of these pairs makes the approach incorrect, that is, we might have to have different fuzzy models for, say, ABOVE and BELOW.

3. A problem that was discussed but not settled is that of quantifying an object to fit into a relational format. We saw in our discussion of Winston's work that simple notions such that as "center of area" or "leftmost" ("topmost," etc.) edge are inadequate. Our fuzzy models will have to look at some complex combination of the properties of the related objects, as in Winston's LEFT OF definition. Limiting our object space as in stage 1 above will hopefully simplify this task.

2 Spatial Relations Among Fuzzy Subsets

2.1 Introduction

Spatial relationships between regions in an image play an important role in scene understanding. Humans are able to quickly ascertain the relationship between two objects, for example "B is to the right of A," or "B is in front of A," but this has turned out to be a somewhat illusive task for automation [7-9].

When the objects in a scene are represented by crisp sets, the all-or-nothing definitions of the subsets actually add to the problem of generating such relational descriptions. It is our belief that definitions of spatial relationships based on fuzzy set theory, coupled with a fuzzy segmentation will yield realistic results.

2.2 The Idea of Projections

This effort presents an initial approach at defining spatial relationships among fuzzy subsets of the image plane.

The idea is to project the fuzzy subsets onto two orthogonal coordinate axes and to utilize fuzzy set theoretic dominance relations to capture the approximate relationships. Similar projections were proposed by Koczy [10].

Let A be a fuzzy subset of an image. Then $A \subseteq U \times V$, where U is the first spatial coordinate axis and V is the second one. In our case, both U and V are subsets of the real numbers (assumed to be the interval [0,1] for convenience). Then $\mu_A(x, y)$ is a 2D fuzzy set in U x V. The projection of A onto U, denoted A_U is that fuzzy subset of U given by $\mu_{Au}(x) = \sup_y \{\mu_A(x,y)\}$ for each $x \in U$. A similar equation defines the projection of A onto V, that is $\mu_{Av}(y) = \sup_x \{\mu_A(x,y)\}$ for each $y \in V$. For a fuzzy subset C of U, the α-cut C^α is defined by $C^\alpha = \{ x \in U \mid \mu_c(x) \geq \alpha \}$ for $\alpha \in [0,1]$. When $\alpha=0$, the inequality is usually considered to be strict and the C^α is called the support of C.

2.3 Definitions of Spatial Relations for Fuzzy Objects

Once the two fuzzy subsets A and B have been projected onto the U and V axes, methods must be defined to access their relative position. In this paragraph we introduce definitions for spatial relations.

Definition 2.1. We say that subset A is to the right of subset B if the projection of A onto the U axis dominates the projection of B, while the projections onto the V axis are (ideally) identical. In other words $\mu_A(\alpha)$ should stay near zero for all α(especially for small α).

Definition 2.2. We say that subset A is to the left of subset B if the projection of B onto the U axis dominates the projection of At while the projections onto the V axis are (ideally) identical. In other words $\mu_v(\alpha)$ should stay near zero for all α (especially for small α).

Definition 2.3. We say that subset A is above subset B if the projection of A onto the V axis dominates the projection of Bt while the projections onto the U axis are (ideally) identical. In other words $\mu_U(\alpha)$ should stay near zero for all α (especially for small α) .

Definition 2.4. We say that subset A is below subset B if the projection of B onto the V axis dominates the projection of At while the projections onto the U axis are (ideally) identical. In other words the grade of membership of an alpha-cut, $\mu_U(\alpha)$, should stay near zero for all α (especially for small α).

If we had a third coordinate axis W, that is, if we considered a three dimensional case, we could define spatial relations "in front of " and "behind." Definitions 2.5 and 2.6 are for these relations. As our concern is in two-dimensional cases we give them here for completeness.

Definition 2.5. We say that subset A is in front of subset B if the projection of A onto the W axis dominates the projection of B, while the projections onto the U axis are (ideally) identical.

In other words $\mu_U(\alpha)$ should stay near zero for all α (especially for small α).

Definition 2.6. We say that subset A is behind subset B if the projection of B onto the W axis dominates the projection of A, while the projections onto the U axis are (ideally) identical. In other words $\mu_U(\alpha)$ should stay near zero for all α (especially for small α).

Definition 2.7. We say that subsets A is inside subset B if the support of A is contained in the support of B.

Definition 2.8. We define that subset A is outside subset B if subset B is inside subset A.

Notice that we can make this assumption as outside relation is symmetric and transitive, and it has a semantic inverse defined above. (We talk about properties of spatial relations in paragraph 2.4.)

2.4   Properties of Spatial Relations for Fuzzy Objects

For the case of interest, let a set X is the set of all fuzzy objects (i.e., fuzzy subsets) of an image. Then, a fuzzy binary relation among the elements of the set X is denoted by R(X,X) or $R(X^2)$ and is a subset of X x X $=X^2$. Various significant types of relations R(X,X) are distinguished on the basis of three different characteristic properties: reflexivity, symmetry, and transitivity. First, let us consider crisp relations.

A crisp relation R(X,X) is reflexive if and only if $(A,A) \in R$ for , every $A \in X$, that is, if every element of X is related to itself. Otherwise, R(X,X) is called irreflexive. If $(A,A) \notin R$ for every $A \in X$, the relation is called anti-reflexive.

A crisp relation R(X,X) is symmetric if and only if for every $(A,B) \in R$, it is also the case that $(B,A) \in R$, where A, B $\in$ X. Thus, whenever an element A is related to an element B through a symmetric relation, then B will also be related to A. If this is not the case for each A, B $\in$ X, then the relation is called asymmetric. If $(A,B) \in R$ and $(B,A) \in R$ implies A= B, then the relation is called anti-symmetric. If either $(B,A) \in R$ or $(A,B) \in R$, whenever A $\neq$ B, then the relation is called strictly anti-symmetric.

A crisp relation R(X,X) is called transitive if and only if (A,C) ∈ R whenever both (A,B) ∈ R and (B,C) ∈ R for at least one B∈ X. In other words, the relation of A and B to C implies the relation of A to C in a transitive relation. A relation that does not satisfy this property is called non-transitive. If (A,C) ∉ R whenever both (A,B) ∈ R and (B,C) ∈ R, then the relation is called anti-transitive.

The properties of reflexivity, symmetry and transitivity can be extended for fuzzy relations R(X,X), by defining them in terms of the membership function of the relation.

Thus R(X,X) is reflexive if and only if $\mu_R(A, A) = 1$ for all A ∈ X. If this is not the case for some A ∈ X, then the relation is called irreflexive; if it is not satisfied for all A ∈ X, the relation is called anti-reflexive. A weaker form of reflexivity, referred to as ε-reflexivity, is sometimes defined by requiring that $\mu_R(A, A) \geq \varepsilon$ where $0 < \varepsilon < 1$.

A fuzzy relation is symmetric if and only if $\mu_R(A,B) = \mu_R(B, A)$ for all A, B ∈ X. Whenever this equality is not satisfied for some A, B ∈ X, the relation is called asymmetric. If the equality fails to hold for all the members of the support of the relation, then the relation is called anti-symmetric, and if it is not satisfied for all A, B ∈ X, then the relation is called strictly anti-symmetric.

A fuzzy relation R(X,X) is transitive (or, more specifically, max-min transitive) if and only if $\mu_R(A,C) \geq \max_{y \in Y} \min [\mu_R (A,B), \mu_R (B,C)]$ is satisfied for each pair (A, C) ∈ X^2. A relation failing to satisfy this inequality for some members of X is called non-transitive, and if $\mu_R(A,C) < \max_{y \in Y} \min [\mu_R (A,B), \mu_R (B,C)]$ for all (A, C) ∈ X^2, then the relation is called anti-transitive. Transitivity of fuzzy relations can be defined in various ways alternative to the form given above. A second common form is known as the max-product transitivity and is defined by $\mu_R(A,C) \geq \max_{y \in Y} [\mu_R (A,B), \mu_R (B,C)]$ for all (A,C) ∈ X^2.

Let us consider the following example. Let R be the fuzzy relation defined on the set of all fuzzy subsets of an image and representing the concept "very near." We may assume that a subset is certainly (i.e., to a degree of 1) very near to itself. The relation is therefore reflexive. Further, if subset A is very near to subset B, then B is certainly very near to A to the same degree. Therefore, the relation is also symmetric. Finally, if subset A is very near to subset B to some degree, say .7, and subset B is very near to subset C to some degree, say .8, it is possible (although not necessary) that subset A is very near to subset C to a smaller degree, say .5. Therefore, the relation is non-transitive.

The definitions 2.1-2.8 are for anti-symmetric and transitive relations, that is, TO THE LEFT (RIGHT) OF, IN FRONT OF (BEHIND), ABOVE (BELOW), INSIDE (OUTSIDE). Each of these relations has a semantic inverse.

After defining the separation measure in the next paragraph, we will be able to define symmetric relations NEAR (FAR). They have semantic inverses, as well. We will also be able to define a reflexive relation TOUCHING.

2.5 Separation Measure

Let A_U, B_U, A_V, B_V be the projections of A and B onto U and V respectively. Since these projections are fuzzy numbers (considering convex objects only), their α-cuts are intervals, i.e.,

$$A_U^\alpha = [A_U^{\alpha l}, A_U^{\alpha r}] \text{, etc.}$$

For the projections of A and B onto the U axis, the α-separation of A and B is defined by [11]:

$$S_U^\alpha = (\overline{A}_U^\alpha - \overline{B}_U^\alpha)^2 / (W_{AU}^\alpha + W_{BU}^\alpha)^2$$

where:

$$\overline{A}_U^\alpha = (A_U^{\alpha l} + A_U^{\alpha r})/2$$

$$\overline{B}_U^\alpha = (B_U^{\alpha l} + B_U^{\alpha r})/2$$

$$W_{AU}^\alpha = (A_U^{\alpha r} - A_U^{\alpha l})/2$$

$$W_{BU}^\alpha = (B_U^{\alpha r} - B_U^{\alpha l})/2.$$

Now, S_U^α is the ratio of the square of difference between the midpoints of the α-cuts and the square of the sum of the half-widths of these intervals. Similar equations are used for the projection of A and B onto the V axis.

Definition 2.9.   We say that A and B are α-separated if $S_U^\alpha > 1$

Definition 2.10. We say that A and B are α-just separated if $S_U^\alpha = 1$

Definition 2.11. We say that A and B are α-overlapping if $S_U^\alpha < 1$.

Theorem 2.1.   i) A_U and B_U are α-separated if and only if

$$A_u^{\alpha r} < B_U^{\alpha l}$$

ii) A_U and B_U are α-just separated if and only if

$$A_u^{\alpha r} = B_U^{\alpha l}$$

iii) A_U and B_U are α-overlapping if and only if

$$A_u^{\alpha r} > B_U^{\alpha l}$$

Proof. The proof follows easily from the geometric interpretation of the numerator of the separation measure given above.

Part ii) — By the definition 2.10, A_U and B_U are α-just separated if $s_U^\alpha = 1$. From the definition of the separation measure we have that:

$$S_U^\alpha = ((A_U^{\alpha 1} + A_U^{\alpha r})/2 - (B_U^{\alpha 1} + B_U^{\alpha r})/2)^2 / ((A_U^{\alpha r} - A_U^{\alpha 1})/2 + (B_U^{\alpha r} - B_U^{\alpha 1})/2)^2$$

The only solution to the quadratic equation $s_U^\alpha = 1$ is $A_U^{\alpha r} = B_U^{\alpha 1}$, which is the desired result. Assuming now that $A_U^{\alpha r} = B_U^{\alpha 1}$, and substituting it into the expression of s_U^α we get that $S_U^\alpha = ((A_U^{\alpha 1} - B_U^{\alpha r})/2)^2 / ((B_U^{\alpha r} - A_U^{\alpha 1})/2)^2$, which concludes the proof.

Part i) — By the definition 2.9, A and B are α-separated $s_U^\alpha > 1$, so we get:

$$((A_U^{\alpha 1} + A_U^{\alpha r})/2 - (B_U^{\alpha 1} + B_U^{\alpha r})/2)^2 / ((A_U^{\alpha r} - A_U^{\alpha 1})/2 + (B_U^{\alpha r} - B_U^{\alpha 1})/2)^2 > 1.$$

Increasing the inequality such that $A_u^{\alpha 1} = A_U^{\alpha r}$ and $B_u^{\alpha 1} = B_U^{\alpha r}$ we have: $\left(A_u^{\alpha 1} - B_U^{\alpha 1}\right)^2 > 0$, which is equivalent to $A_u^{\alpha 1} < B_U^{\alpha 1}$ or $A_u^{\alpha 1} > B_U^{\alpha 1}$. We can assume without no loss of generality that $A_u^{\alpha 1} < B_U^{\alpha 1}$, which is the expected inequality. Now, assuming that $A_u^{\alpha r} < B_U^{\alpha 1}$ and substituting it into the expression of separation measure, we have that $s_U^\alpha > 1$, which by definition 2.9 means that A_U and B_U are α-separated, and which completes the proof.

Part iii) — By the definition 2.11, A_U and B_U are α-overlapping if $s_U^\alpha < 1$, so:

$$((A_U^{\alpha 1} + A_U^{\alpha r})/2 - (B_U^{\alpha 1} + B_U^{\alpha r})/2)^2 / ((A_U^{\alpha r} - A_U^{\alpha 1})/2 + (B_U^{\alpha r} - B_U^{\alpha 1})/2)^2 < 1.$$

Increasing the inequality such that $A_u^{\alpha 1} = B_U^{\alpha 1}$ and $A_u^{\alpha r} = B_U^{\alpha r}$ we get: $(A_u^{\alpha r} - B_u^{\alpha 1})^2 > 0$ is equivalent to $A_u^{\alpha r} > B_u^{\alpha 1}$ or $A_u^{\alpha r} < B_u^{\alpha 1}$. Again, if A is the first set from the left, we get that $A_u^{\alpha r} > B_u^{\alpha 1}$, which is the desired inequality. Now assuming that $A_u^{\alpha r} > B_u^{\alpha 1}$ and substituting it into the expression of separation measure, we have that $s_U^\alpha < 1$, which by definition 2.11 means that A_U and B_U are α-overlapping, and which concludes the proof.

The value of these definitions and theorems is twofold. They incorporate the fuzziness in the description of image regions, i.e., they use fuzzy subsets of the plane. Secondly, they deal with the ambiguity of defining spatial relationships in the plane. By this we mean that it is possible that parts of the two II sets can overlap (small α) and yet be well separated for large α.

The values of S_U^α can get arbitrarily large as the widths of the level set intervals get small. In order to create a fuzzy membership function, we will map the interval $[0,\infty)$ into $[0, 1]$ by an "S-shaped function" [11] as follows. For a given α, suppose $A_U^\alpha=[0,0.2]$ and $B_U^\alpha=[0.8,1]$ (recall that we have scaled the domain of the image into the unit square), then $S_U^\alpha = 16$. This amount of separation (or more) will be considered complete, i.e., $\mu(S_U^\alpha) = 1$ if $S_U^\alpha \geq 16$. Also we will require that $\mu(0) = 0$, $\mu(1) = 0.5$, and $\mu(16) = 0$. Such a function is defined in our case by:

$$
\mu(S) = \left.\begin{cases}
0.5\ S^2 \\[1em]
-0.0022\ S^2 + 0.0711S + 0.4311 \\[1em]
1
\end{cases}\right\}
\qquad
\begin{array}{l}
0 \leq S \leq 1 \\[1em]
1 < S \leq 16 \\[1em]
S > 16
\end{array}
$$

We can now define the TOUCHING relation.

Definition 2.12. We define that subset A^α is touching subset B^α if the separation measure between subsets A^α and B^α equals 1 in one projection, and equals 0 in the other one, that is $\mu(S_U^\alpha) = 0$ and $\mu(S_U^\alpha) = 0.5$ (or vice versa).

Definition 2.13. We define that subset A^α is near subset B^α if $0.5 < \mu(S_U^\alpha) \leq 0.75$. (Similarly for the projections onto the V axis.)

Definition 2.14. We define that subset B^α is far from subset A^α if $\mu(S_U^\alpha) > 0.75$. (Similarly for the projections onto the V axis.)

It should be mentioned that a similar idea was explored by Koczy [10], however neither separation measure nor a function of it was considered there.

2.6 The Model for Spatial Relationships

The model for given spatial relationships can now be defined from the fuzzy subsets μ_U and μ_V of $[0, 1]$. For example, to model the relationship "A IS To THE RIGHT OF B," we would like the projection of A onto the U axis to dominate that of B; whereas the projections should (ideally) be identical on the V axis. That is, $\mu_V(\alpha)$ should stay near zero for all α (especially for small α). Similar observations can be made for "ABOVE," "BELOW."

Instead of dealing with two fuzzy subsets, μ_U and μ_V can be combined into a single set from which the relationship can be determined. Fuzzy set theory offers an infinite number of aggregation operators, which, given two pieces of evidence (values in $[0, 1]$) can produce essentially any composite value between 0 and 1, depending on the type of connective and the parameters chosen. Union operators produce values greater than or equal to the maximum of the two numbers; intersection operators give a result less than or equal to the minimum; and generalized means fill the gap between the minimum and maximum [12].

"TO THE RIGHT OF" should therefore be a combination of μ_U and the complement of μ_V since its large values signify that the level sets of A are "above or below" those of B. For the experiments described in paragraph 2.7, we chose a generalized mean

$$m(\mu_U, \mu_v) = [W\mu_U^p + (1-W)(1-\mu_v)^p)]^{1/p}$$

as the aggregation connective [12]. In this way, we can put higher weight on the horizontal component with decreased compensation as the level sets diverge vertically. Note that if P→∞, then we have [12]:

$$\lim_{p \to \infty} m(\mu_U, \mu_v) = \max(\mu_U, \mu_v)$$

Either the two fuzzy sets μ_U and (1-μ_V) or the single aggregated set m(μ_U,μ_V) can be used to define the relation "A IS TO RIGHT OF B." If a single value for the degree to which the two sets satisfy the relation is desired, we can construct a measure from the sets -such as the integral of the fuzzy number, or the output of an Ordered Weighted Average (OWA) [13]. An alternate approach is to use the curves directly to define a linguistic assessment of the relation. Here, it is necessary to define fuzzy sets representing terms used in the relation, such as "to the right of," "somewhat to the right of," "barely to the right of," "very to the right of," etc. These sets could be defined by the designer of the system, or perhaps, by utilizing a group of humans to give relative comparisons of a set of examples. The actual curve is then matched to the closest term available to give the linguistic assessment. This process is known as linguistic approximation [14].

2.7   Results of Sample Systems

All the definitions and theorem listed above were tested using simulated data on a Sun Workstation in the Computer Vision Laboratory at the University of Missouri-Columbia. We worked with fuzzy subsets with two-sided drum like shaped membership functions on projections. The experiments were as follows. Let us consider an image containing two fuzzy subsets A and B whose membership functions are identical Gaussians, but with different mean locations. The set B will be fixed with mean (0.5, 0.5). Table 1 shows the fuzzy set μ_V generated from eight choices of locations for the mean of A (assume that the V coordinate for the mean is 0.5). As can be seen, as the set A moves to the right, the fuzzy set μ_U increases for all α. Recall that the value μ_U (α) = 0.5 represents the just separated condition. Three of these sets are shown graphically in Fig. 4.

The seven α-values are 0.011, 0.135, 0.258, 0.606, 0.796, 0.882, 0.923. They were chosen to get the following ranges from the mean of Gaussian functions: ±0.4σ, ±0.5σ, ±0.6745σ, ±σ, ±1.645σ, ±2σ, ±3σ, where σ is a standard deviation. Since the projections onto V for these sets are the same as the projections onto U, these fuzzy sets in Table 1 can be used to simulate other placing of A relative to B, e.g., to the northeast or southeast. Table 2 shows four cases for the placement of the center of set A along with the aggregated fuzzy set generated from both projections. We used the generalized mean with W= 0.75 and P=2. The first case represents a set A, which is due east of B. Here the combined values are larger than those for the U projections only. In fact, even the smallest α (0.011) gives rise to a membership larger than 0.5 (the just separated crossover point). In case 2, the set A has moved to the north east of B. The movement north effectively decreases the membership in the fuzzy set "A is to the right of B." Cases 3 and 4 depict the situation where A is directly above B. As the centers move farther apart, the membership drops dramatically.

If we change either the weight W or the exponent P, we can alter the shape of the resultant fuzzy set. Tables 3 and 4 illustrate this. We used the following parameters: W= 0.8 and P= 3 to get the combined membership function in the Table 3, and W=0.9, P=4 to get the membership function in the Table 4. The shapes of the resultant fuzzy sets have been changed, but still, even for the smallest α, we get memberships greater than 0.5 (the just separated crossover point). Also, switching to a union or intersection operator would have an even more dramatic effect on the consequent curve. This flexibility can be used to model the degree of optimism or pessimism that is desired in satisfying the relation "to the right of." For the relation "to the left of " we have that as the set A moves to the left, the fuzzy set μ_V increases for all α. Three of these sets are shown graphically in Fig. 5. We can notice that movement north of the set A decreases the membership in the fuzzy set "A is to the left of B." For the relation "above," the movement up of the set A results in increasing the fuzzy set μ_V for all α. Four of these sets are shown graphically in Fig. 6. As the set A moves down from the set B, the fuzzy set μ_V increases for all α. Four of these sets are shown again in Fig. 7. We can also notice that movement east of the set A effectively decreases the membership in the fuzzy set "A is below B." As the sets move farther apart, the membership drops rapidly. Again, for different parameters W and P, which results in altering the resultant fuzzy sets, even the smallest a gives rise to a membership larger than 0.5 (for all the spatial relations. Figs. 4-7 depict the dominance membership functions calculated from the projections of B and various A's onto the U or V axes for the various spatial relations.

2.8 Conclusions

An approach, based on the concept of dominance in fuzzy set theory, for modeling spatial relationships among fuzzy subsets of an image has been presented. Simulation results were presented to corroborate the theory and demonstrate the utility of the approach to the image description.

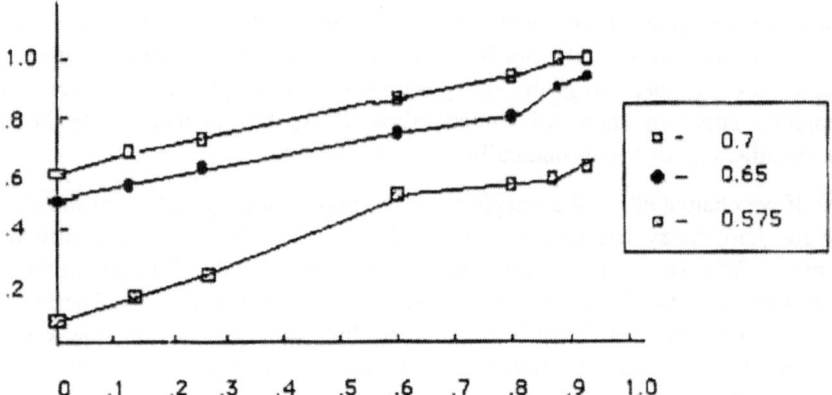

Fig. 4. Dominance membership functions calculated from the projections of subsets B and A onto the U axis for the relation "to the right of."

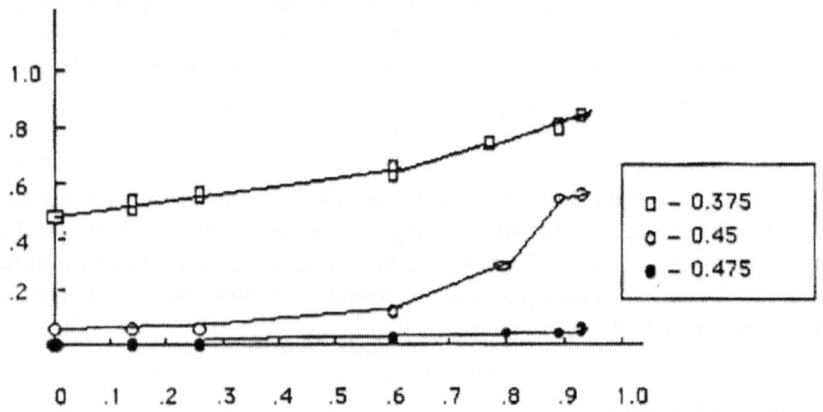

Fig. 5. Dominance membership functions calculated from the projections of subsets B and A onto the U axis for the relation "to the left of."

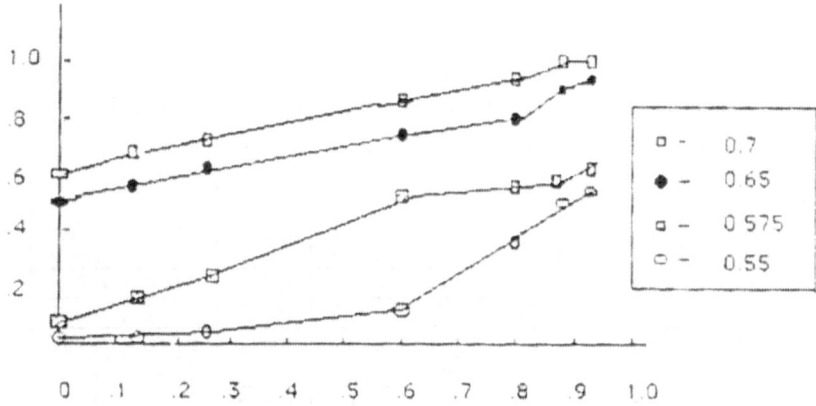

Fig. 6. Dominance membership functions calculated from the projections of subsets B and A onto the W axis for the relation "above."

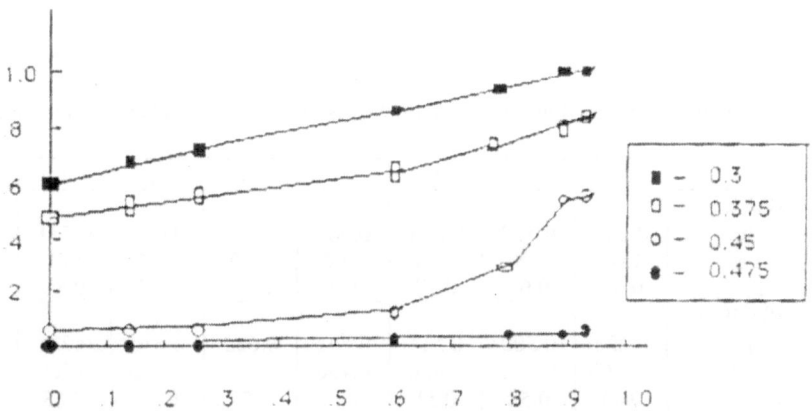

Fig. 7. Dominance membership functions calculated from the projections of subsets B and A onto the W axis for the relation "below."

TABLE 1

Fuzzy membership functions generated from the projection of A onto U axis.

Mean of Projections of A	α_1	α_2	α_3	α_4	α_5	α_6	α_7
0.525	0.001	0.002	0.003	0.008	0.017	0.031	0.049
0.550	0.014	0.031	0.046	0.125	0.275	0.500	0.516
0.575	0.070	0.158	0.234	0.508	0.543	0.579	0.623
0.600	0.222	0.500	0.514	0.564	0.622	0.680	0.731
0.625	0.503	0.536	0.558	0.631	0.712	0.788	0.851
0.650	0.532	0.580	0.609	0.706	0.806	0.891	0.950
0.675	0.567	0.628	0.665	0.783	0.893	0.968	0.998
0.700	0.604	0.680	0.724	0.857	0.962	1.00	1.00

TABLE 2

Combined membership function for the relation "A is to the right of B" (W= 0.75, P= 2).

	α_1	α_2	α_3	α_4	α_5	α_6	α_7
(0.6,0.5)							
μ_U	0.222	0.500	0.514	0.564	0.622	0.680	0.731
$1-\mu_V$	1.00	1.00	1.00	1.00	1.00	1.00	1.00
$m(\mu_U, \mu_V)$	0.54	0.66	0.67	0.70	0.73	0.77	0.81
(0.6,0.6)							
μ_U	0.222	0.500	0.514	0.564	0.622	0.680	0.731
$1-\mu_V$	0.778	0.500	0.486	0.436	0.378	0.320	0.269
$m(\mu_U, \mu_V)$	0.43	0.50	0.51	0.53	0.57	0.61	0.65
(0.5,0.6)							
μ_U	0.00	0.00	0.00	0.00	0.00	0.00	0.00
$1-\mu_V$	0.778	0.500	0.486	0.436	0.378	0.320	0.269
$m(\mu_U, \mu_V)$	0.39	0.25	0.24	0.22	0.19	0.16	0.13
(0.5,0.7)							
μ_U	0.00	0.00	0.00	0.00	0.00	0.00	0.00
$1-\mu_V$	0.396	0.320	0.276	0.143	0.038	0.00	0.00
$m(\mu_U, \mu_V)$	0.20	0.16	0.14	0.07	0.02	0.00	0.00

TABLE 3

Combined membership function for the relation "A is to the right of B" (W=0.8, P=3).

	α_1	α_2	α_3	α_4	α_5	α_6	α_7
(0.6,0.5)							
μ_U	0.222	0.500	0.514	0.564	0.622	0.680	0.731
$1-\mu_V$	1.00	1.00	1.00	1.00	1.00	1.00	1.00
$m(\mu_U, \mu_V)$	0.59	0.67	0.68	0.70	0.73	0.77	0.81
(0.6,0.6)							
μ_U	0.222	0.500	0.514	0.564	0.622	0.680	0.731
$1-\mu_V$	0.778	0.500	0.486	0.436	0.378	0.320	0.269
$m(\mu_U, \mu_V)$	0.47	0.50	0.51	0.54	0.59	0.64	0.68
(0.5,0.6)							
μ_U	0.00	0.00	0.00	0.00	0.00	0.00	0.00
$1-\mu_V$	0.778	0.500	0.486	0.436	0.378	0.320	0.269
$m(\mu_U, \mu_V)$	0.45	0.29	0.28	0.25	0.22	0.18	0.15
(0.5,0.7)							
μ_U	0.00	0.00	0.00	0.00	0.00	0.00	0.00
$1-\mu_V$	0.396	0.320	0.276	0.143	0.038	0.00	0.00
$m(\mu_U, \mu_V)$	0.23	0.18	0.16	0.08	0.02	0.00	0.00

TABLE 4

Combined membership function for the relation "A is to the right of B" (W=0.9, P=4).

	α_1	α_2	α_3	α_4	α_5	α_6	α_7
(0.6,0.5)							
μ_U	0.222	0.500	0.514	0.564	0.622	0.680	0.731
$1-\mu_V$	1.00	1.00	1.00	1.00	1.00	1.00	1.00
$m(\mu_U, \mu_V)$	0.56	0.63	0.64	0.66	0.70	0.74	0.79
(0.6,0.6)							
μ_U	0.222	0.500	0.514	0.564	0.622	0.680	0.731
$1-\mu_V$	0.778	0.500	0.486	0.436	0.378	0.320	0.269
$m(\mu_U, \mu_V)$	0.44	0.50	0.51	0.55	0.61	0.66	0.71
(0.5,0.6)							
μ_U	0.00	0.00	0.00	0.00	0.00	0.00	0.00
$1-\mu_V$	0.778	0.500	0.486	0.436	0.378	0.320	0.269
$m(\mu_U, \mu_V)$	0.43	0.28	0.27	0.24	0.21	0.18	0.15
(0.5,0.7)							
μ_U	0.00	0.00	0.00	0.00	0.00	0.00	0.00
$1-\mu_V$	0.396	0.320	0.276	0.143	0.038	0.00	0.00
$m(\mu_U, \mu_V)$	0.22	0.18	0.15	0.08	0.02	0.00	0.00

References

1. J. Freeman. The Modeling of Spatial Relations. *Computer Graphics and Image Processing*, 4:156-171, 1975.
2. P. Winston. *The Psychology of Computer Vision*, McGraw Hill, New York , 1975.
3. L. Urdang and M. Manser. *The Pan Dictionary of Synonyms and Antonyms*, Pan Books, London, Sydney and Auckland, 1980.
4. T. Givon. Notes on the Semantic Structure of English Adjectives. *Language*, 46:816-837, 1971.
5. H. H. Clark. Comprehending Comparatives. In *Advances in Psycholinguistics*, G. F. d'Arcais and W. J. M. Levett (Eds.), North-Holland, Amsterdam, 1970.
6. H. H. Clark and W. G. Chase. On the Process of Comparing Sentences Against Pictures. *Cognitive Psych.*, 3:472-517, 1972.
7. K. P. Gapp. Basic Meanings of Spatial Relations: Computation and Evaluation in 3D Space. In *AAAI-94*, pages 1393-1398, Seattle, WA, 1994.
8. I. Bloch. Fuzzy Relative Position between Objects in Image Processing: Morphological Approach. *IEEE Transactions on Pattern Analysis and Machine Intelligence*, 7(21):657-664, 1999.
9. J. Keller and X. Wang. A Fuzzy Rule-based Approach for Scene Description Involving Spatial Relationships. *Computer Vision and Image Understanding*, 80:21-41, 2000.
10. L. T. Koczy. On the Description of Relative Position of Fuzzy Patterns. *Pattern Recognition Letters*, 8:21-28, 1981.
11. J. Keller and L. Sztandera. Spatial Relations Among Fuzzy Subsets of an Image. In *1st Int. Symposium on Uncertainty Modeling and Analysis*, pages 207-211, College Park, MD, 1990.
12. G. J. Klir and T. Folger. *Fuzzy Sets, Uncertainty, and Information*, Prentice Hall, 1988.
13. R. R. Yager. On Ordered Weighted Averaging Aggregation Operations in Multicriteria Decision Making. *IEEE Transactions on Systems, Man and Cybernetics*, 18(1):183-190, 1988.
14. L. A. Zadeh. The Concept of a Linguistic Variable and Its Application in Approximate Reasoning. *Information Sciences*, 8:301-357, 1975.

Mathematical Morphology and Spatial Relationships: Quantitative, Semi-Quantitative and Symbolic Settings

Isabelle Bloch

Ecole Nationale Supérieure des Télécommunications
Département TSI - CNRS URA 820
46 rue Barrault, 75013 Paris, France
Isabelle.Bloch@enst.fr

Abstract. Basic mathematical morphology operations rely mainly on local information, based on the concept of structuring element. But mathematical morphology also deals with more global and structural information since several spatial relationships can be expressed in terms of morphological operations (mainly dilations). The aim of this paper is to show that this framework allows to represent in a unified way spatial relationships in various settings: a purely quantitative one if objects are precisely defined, a semi-quantitative one if objects are imprecise and represented as spatial fuzzy sets, and a qualitative one, for reasoning in a logical framework about space. This is made possible thanks to the strong algebraic structure of mathematical morphology, that finds equivalents in set theoretical terms, fuzzy operations and logical expressions.

1 Introduction

Mathematical morphology is originally based on set theory. It has been introduced in 1964 by Matheron [37,38], in order to study porous media. But this theory evolved rapidly to a general theory of shape and its transformations, and was applied in particular in image processing and pattern recognition [50,51]. Additionally to its set theoretical foundations, it also relies on topology on sets, on random sets, on topological algebra, on integral geometry, on lattice theory.

Its basic transformations rely mainly on local information, based on the concept of structuring element. But mathematical morphology also deals with more global and structural information since several spatial relationships can be expressed in terms of morphological operations (mainly dilations).

The aim of this paper is to show that this framework allows to represent in a unified way spatial relationships in various settings: a purely quantitative one if objects are precisely defined, a semi-quantitative one if objects are imprecise and represented as spatial fuzzy sets, and a qualitative one, for reasoning in a logical framework about space. This is made possible thanks to the strong algebraic structure of mathematical morphology, that finds equivalents in set theoretical terms, fuzzy operations and logical expressions.

The interest of relationships between objects has been highlighted in very different types of works: in vision, for identifying shapes and objects, in database system management, for supporting spatial data and queries, in artificial intelligence, for planning and reasoning about spatial properties of objects, in cognitive and perceptual psychology, in geography, for geographic information systems. According to the semantical hierarchy proposed in [34], we consider topological and metric relationships (corresponding to levels 3 and 4 of this hierarchy). Many authors have stressed the importance of topological relationships (which include part-whole relationships such as inclusion, exclusion, adjacency, etc.), e.g. [1,54,45,16,2,15,33,44]. But distances and directional relative position (constituting the metric relationships) are also important, e.g. [43,22,34,24,32,35].

Definitions of basic morphological operations and of its extensions to fuzzy sets and to logical expressions are given in Section 2. We restrict our presentation to operations based on structuring elements. More general operators can be defined in the complete lattice framework based on the notion of adjunction (see e.g. [28,18]). If the operations satisfy an additional property of invariance under translation (in the spatial domain), then there exists a structuring element B such that they take the forms used here [50,51]. This property is often a requirement when dealing with spatial information, and therefore we do not consider the more general framework. Section 2 contains all necessary theoretical background for the three following sections.

We address three questions in this paper, which govern its structure. In Section 3, we consider the problem of defining and computing spatial relationships between two objects, in both the crisp and fuzzy cases.

Then in Section 4, we propose a way to represent spatial knowledge in the spatial domain. Given a reference object, we define a spatial fuzzy set that represents the area of the space where some relationship to this reference object is satisfied (to some degree). The advantage of these representations is that they map all types of spatial knowledge in the same space, which allows for their fusion and for spatial reasoning.

Finally in Section 5 we show that spatial relationships can be expressed in the framework of normal modal logics, using morphological operations applied on logical formulas. This can be useful for symbolic (purely qualitative) spatial reasoning.

2   Basic Morphological Operations, Fuzzy and Logical Extensions

2.1   Classical Morphology on Sets and Functions

Let us first recall the definitions of dilation and erosion of a set X by a structuring element B in a space S (e.g. $\mathbb{R}^n$, or $\mathbb{Z}^n$ for discrete spaces like images), denoted respectively by $D_B(X)$ and $E_B(X)$ [50]:

$$D_B(X) = \{x \in S \mid B_x \cap X \neq \emptyset\}, \tag{1}$$

$$E_B(X) = \{x \in \mathcal{S} \mid B_x \subset X\}, \tag{2}$$

where B_x denotes the translation of B at point x.

In these equations, B defines a neighborhood that is considered at each point. It can also be seen as a relationship between points.

These definitions extend to functions f and to functional structuring elements [50].

From these two fundamental operations, a lot of others can be built [50,51]. Of particular interest are opening and closing, defined as:

$$O_B(X) = D_{\check{B}}[E_B(X)], \tag{3}$$

$$C_B(X) = E_{\check{B}}[D_B(X)], \tag{4}$$

where $\check{B}$ denotes the symmetrical of B with respect to the origin of $\mathcal{S}$.

2.2 Fuzzy Mathematical Morphology

Several definitions of mathematical morphology on fuzzy sets with fuzzy structuring elements have been proposed in the literature (see e.g. [10,52,17]). Here we use the approach of [10] using t-norms and t-conorms as fuzzy intersection and fuzzy union[1]. However, what follows applies as well if other definitions are used. Erosion of a fuzzy set μ by a fuzzy structuring element ν, both defined in a space $\mathcal{S}$ (for instance $\mathcal{S} = \mathbb{R}^n$), is defined as:

$$\forall x \in \mathcal{S}, \; E_\nu(\mu)(x) = \inf_{y \in \mathcal{S}} T[c(\nu(y - x)), \mu(y)], \tag{5}$$

where T is a T-conorm and c a fuzzy complementation. Fuzzy dilation is defined as:

$$\forall x \in \mathcal{S}, \; D_\nu(\mu)(x) = \sup_{y \in \mathcal{S}} t[\nu(y - x), \mu(y)], \tag{6}$$

where t is the t-norm associated to the t-conorm T with respect to the complementation c .

Opening and closing are defined as a combination of dilation and erosion as in the crisp case.

These definitions guarantee that most properties are preserved when extended to fuzzy sets. Extensivity of closing, anti-extensivity of opening, and idempotence of these operations are satisfied only for specific t-norms and t-conorms (see [10] for further details, and proofs in [5]). We summarize here the main properties:

- erosion and dilation (respectively opening and closing) are dual with respect to the complementation c;

---

[1] A triangular norm (or t-norm) is a function from $[0, 1] \times [0, 1]$ into $[0, 1]$ which is commutative, associative, increasing, and for which 1 is unit element and 0 is null element. Examples of t-norms are min, product, etc. [19].

- if the structuring element is binary, the same definitions as in classical mathematical morphology are obtained;
- compatibility with translations;
- local knowledge property;
- continuity if the t-norm is continuous (which is most often the case);
- increasingness of all operations with respect to inclusion;
- extensivity of dilation and anti-extensivity of erosion iff $\nu(0) = 1$ (this corresponds to the condition that the origin should belong to the structuring element in the crisp case);
- extensivity of closing, anti-extensivity of opening and idempotence of these two operations if and only if $t[b, u(c(b), a)] \leq a$, which is satisfied for Lukasiewicz t-norm and t-conorm (note that this property always holds for definitions based on residual implications, as proposed in [17]);
- commutation of dilation with union (with intersection for erosion);
- iteration property of dilation.

2.3   Morpho-Logics

In this Section, we express morphological operations in a symbolic framework, using logical formulas. Let us first introduce some notations. Let PS be a finite set of propositional symbols. The language is generated by PS and the usual connectives, to which we will add modal operators in the following. Well-formed formulas will be denoted by Greek letters φ, ψ... Kripke's semantics is used. Worlds will be denoted by ω, ω' and the set of all worlds by Ω. $Mod(\varphi) = \{\omega \in \Omega \mid \omega \models \varphi\}$ is the set of all worlds where φ is satisfied.

The underlying idea for constructing morphological operations on logical formulas (as presented in [9]) is to consider set interpretations of formulas and worlds. Since in classical propositional logics, the set of formulas is isomorphic to 2^Ω, i.e., knowing a formula is equivalent to knowing the set of worlds where the formula is satisfied, we can identify φ with $Mod(\varphi)$, and then apply set-theoretic morphological operations. We recall that $Mod(\varphi \vee \psi) = Mod(\varphi) \cup Mod(\psi)$, $Mod(\varphi \wedge \psi) = Mod(\varphi) \cap Mod(\psi)$, and $Mod(\varphi) \subset Mod(\psi)$ iff $\varphi \rightarrow \psi$.

Using the previous equivalences, and based on set definitions of morphological operators [50], dilation and erosion of a formula φ have been defined in [9] as follows:

$$Mod(D_B(\varphi)) = \{\omega \in \Omega \mid B(\omega) \cap Mod(\varphi) \neq \emptyset\}, \tag{7}$$

$$Mod(E_B(\varphi)) = \{\omega \in \Omega \mid B(\omega) \models \varphi\}. \tag{8}$$

In these equations, the structuring element B represents a relationship between worlds, i.e. $\omega' \in B(\omega)$ iff ω' satisfies some relationship with ω. The condition in Equation 7 expresses that the set of worlds in relation to ω should be consistent with φ, i.e.:

$$\exists \omega' \in B(\omega), \ \omega' \models \varphi.$$

The condition in Equation 8 is stronger and expresses that φ should be satisfied in all worlds in relation to ω.

The structuring element B representing a relationship between worlds defines a "neighborhood" of worlds. If it is symmetrical, it leads to symmetrical structuring elements. If it is reflexive, it leads to structuring elements such that $\omega \in B_\omega$, which leads to interesting properties, as will be seen later.

An interesting way to choose the relationship is to base it on distances between worlds, which is an important information in spatial reasoning. This allows to define sequences of increasing structuring elements defined as the balls of a distance. For any distance δ between worlds, a structuring element of size n centered at ω takes the following form:

$$B^n(\omega) = \{\omega' \in \Omega \mid \delta(\omega, \omega') \leq n\}. \tag{9}$$

For instance a distance equal to 1 can represent a connectivity relation between worlds, defined for instance as a difference of one literal (i.e. one literal instantiated differently in both worlds).

Now we consider the framework of normal modal logics [29,14] and use an accessibility relation as relation between worlds. We define an accessibility relation from any structuring element B as follows:

$$R(\omega, \omega') \text{ iff } \omega' \in B(\omega). \tag{10}$$

Conversely, a structuring element can be defined from an accessibility relation.

The accessibility relation R is reflexive iff $\forall \omega \in \Omega$, $\omega \in B(\omega)$. It is symmetrical iff $\forall (\omega, \omega') \in \Omega^2$, $\omega \in B(\omega') \Leftrightarrow \omega' \in B(\omega)$. In the following we will restrict to symmetrical relations. In general, accessibility relations derived from a structuring element are not transitive.

Let us now consider the two modal operators $\Box$ and $\Diamond$ defined from the accessibility relation as [14]:

$$\mathcal{M}, \omega \models \Box\varphi \text{ iff } \forall \omega' \in \Omega, \ R(\omega, \omega') \Rightarrow \mathcal{M}, \omega' \models \varphi, \tag{11}$$

$$\mathcal{M}, \omega \models \Diamond\varphi \text{ iff } \exists \omega' \in \Omega, \ R(\omega, \omega') \text{ and } \mathcal{M}, \omega' \models \varphi, \tag{12}$$

where $\mathcal{M}$ denotes a standard model related to R, that will be skipped in the notations in the following (it will be always implicitly related to the considered accessibility relation).

Equation 11 can be rewritten as:

$$\omega \models \Box\varphi \Leftrightarrow \{\omega' \in \Omega \mid R(\omega, \omega')\} \models \varphi$$
$$\Leftrightarrow \{\omega' \in \Omega \mid \omega' \in B(\omega)\} \models \varphi$$
$$\Leftrightarrow B(\omega) \models \varphi,$$

which exactly corresponds to the definition of erosion of a formula according to Equation 8.

In a similar way, Equation 12 can be rewritten as:

$$\omega \models \Diamond\varphi \Leftrightarrow \{\omega' \in \Omega \mid R(\omega, \omega')\} \cap Mod(\varphi) \neq \emptyset$$
$$\Leftrightarrow \{\omega' \in \Omega \mid \omega' \in B(\omega)\} \cap Mod(\varphi) \neq \emptyset$$
$$\Leftrightarrow B(\omega) \cap Mod(\varphi) \neq \emptyset,$$

which exactly corresponds to a dilation according to Equation 7.

This shows that we can define modal operators derived from an accessibility relation as erosion and dilation with a structuring element:

$$\Box\varphi = E_B(\varphi), \tag{13}$$

$$\Diamond\varphi = D_B(\varphi). \tag{14}$$

Theorem 1 *The modal logic constructed from erosion and dilation has the following theorems and rules of inference*[2]:

- **T**: $\Box\varphi \to \varphi$ *and* $\varphi \to \Diamond\varphi$ *(if B is such that* $\forall \omega \in \Omega, \omega \in B(\omega)$, *leading to a reflexive accessibility relation).*
- **Df**: $\Diamond\varphi \leftrightarrow \neg\Box\neg\varphi$ *and* $\Box\varphi \leftrightarrow \neg\Diamond\neg\varphi$.
- **D**: $\Box\varphi \to \Diamond\varphi$.
- **B**: $\Diamond\Box\varphi \to \varphi$ *and* $\varphi \to \Box\Diamond\varphi$.
- **5c**: $\Box\Diamond\varphi \to \Diamond\varphi$ *and* $\Box\varphi \to \Diamond\Box\varphi$ *(if B is such that* $\forall \omega \in \Omega, \omega \in B(\omega)$*).*
- **4c**: $\Box\Box\varphi \to \Box\varphi$ *and* $\Diamond\varphi \to \Diamond\Diamond\varphi$ *(if B is such that* $\forall \omega \in \Omega, \omega \in B(\omega)$*).*
- **N**: $\Box\top$ *and* $\neg\Diamond\bot$.
- **M**: $\Box(\varphi \wedge \psi) \to (\Box\varphi \wedge \Box\psi)$ *and* $(\Diamond\varphi \vee \Diamond\psi) \to \Diamond(\varphi \vee \psi)$.
- **M'**: $\Diamond(\varphi \wedge \psi) \to (\Diamond\varphi \wedge \Diamond\psi)$ *and* $(\Box\varphi \vee \Box\psi) \to \Box(\varphi \vee \psi)$.
- **C**: $(\Box\varphi \wedge \Box\psi) \to \Box(\varphi \wedge \psi)$ *and* $\Diamond(\varphi \vee \psi) \to (\Diamond\varphi \vee \Diamond\psi)$.
- **R**: $(\Box\varphi \wedge \Box\psi) \leftrightarrow \Box(\varphi \wedge \psi)$ *and* $\Diamond(\varphi \vee \psi) \leftrightarrow (\Diamond\varphi \vee \Diamond\psi)$.
- **RN**:

$$\frac{\varphi}{\Box\varphi}.$$

- **RM**:

$$\frac{\varphi \to \psi}{\Box\varphi \to \Box\psi} \ and \ \frac{\varphi \to \psi}{\Diamond\varphi \to \Diamond\psi}.$$

- **RR**:

$$\frac{(\varphi \wedge \varphi') \to \psi}{(\Box\varphi \wedge \Box\varphi') \to \Box\psi} \ and \ \frac{(\varphi \vee \varphi') \to \psi}{(\Diamond\varphi \vee \Diamond\varphi') \to \Diamond\psi}.$$

- **RE**:

$$\frac{\varphi \leftrightarrow \psi}{\Box\varphi \leftrightarrow \Box\psi} \ and \ \frac{\varphi \leftrightarrow \psi}{\Diamond\varphi \leftrightarrow \Diamond\psi}.$$

- **K**: $\Box(\varphi \to \psi) \to (\Box\varphi \to \Box\psi)$ *and by duality* $(\neg\Diamond\varphi \wedge \Diamond\psi) \to \Diamond(\neg\varphi \wedge \psi)$.

---

[2] We use similar notations as in [14] for these theorems and rules of inference.

Let us now denote by $\square^n$ the iteration of n times $\square$ (i.e. n erosions by the same structuring element). Since the succession of n erosions by a structuring element is equivalent to one erosion by a larger structuring element, of size n (iterativity property of erosion), $\square^n$ is a new modal operator, constructed as in Equation 13. In a similar way, we denote by $\Diamond^n$ the iteration of n times $\Diamond$, which is again a new modal operator, due to iterativity property of dilation, constructed as in Equation 14 with a structuring element of size n. We set $\square^1 = \square$ and $\Diamond^1 = \Diamond$.

Theorem 2 *We have the additional following theorems:*

- $\square^n \square^{n'} \varphi \leftrightarrow \square^{n+n'} \varphi$, *and* $\Diamond^n \Diamond^{n'} \varphi \leftrightarrow \Diamond^{n+n'} \varphi$ *(iterativity properties of dilation and erosion).*
- $\Diamond \square \Diamond \square \varphi \leftrightarrow \Diamond \square \varphi$, *and* $\square \Diamond \square \Diamond \varphi \leftrightarrow \square \Diamond \varphi$ *(idempotence of opening and closing).*
- *More generally, from properties of closing and opening:*

$$\Diamond^n \square^n \Diamond^{n'} \square^{n'} \varphi \leftrightarrow \Diamond^{n'} \square^{n'} \Diamond^n \square^n \varphi \leftrightarrow \Diamond^{\max(n,n')} \square^{\max(n,n')} \varphi,$$

and

$$\square^n \Diamond^n \square^{n'} \Diamond^{n'} \varphi \leftrightarrow \square^{n'} \Diamond^{n'} \square^n \Diamond^n \varphi \leftrightarrow \square^{\max(n,n')} \Diamond^{\max(n,n')} \varphi.$$

- *For* $n < n'$, $\Diamond^n \varphi \rightarrow \Diamond^{n'} \varphi$, $\square^{n'} \varphi \rightarrow \square^n \varphi$, $\square^n \Diamond^n \varphi \rightarrow \square^{n'} \Diamond^{n'} \varphi$, $\Diamond^{n'} \square^{n'} \varphi \rightarrow \Diamond^n \square^n \varphi$.

All these definitions and properties extend to the fuzzy case, if we consider fuzzy formulas, i.e. formulas φ for which $Mod(\varphi)$ is a fuzzy set of Ω. The fuzzy structuring element can be interpreted as a fuzzy relation between worlds.

The use of fuzzy structuring elements will appear as particularly useful for expressing intrinsically vague spatial relationships such as directional relative position.

3   Computing Spatial Relationships from Mathematical Morphology: Quantitative and Semi-Quantitative Setting

In this Section we consider the problem of defining and computing spatial relationships between two objects. This problem occurs for instance in model-based pattern recognition, where objects in a scene can be recognized and labeled according to a model of this scene based on their own characteristics but also on relationships linking them to each other. Relationships are particularly useful when information about the objects (like shape) is not reliable or subject to variability. This occurs for instance if the model is schematic, like a map or an anatomical atlas.

We consider the general case of a 3D space S, where objects can have any shape and any topology, and consider both topological and metric relationships [34]. We distinguish also between relationships that are mathematically well defined and relationships that are intrinsically vague. Topological relationships (such as set relationships and adjacency) and distances belong to the first class. If the objects are precisely defined, their relationships can be defined and computed in a numerical (purely quantitative) setting. But if the objects are imprecise, as is often the case if they are extracted from images, then the semi-quantitative framework of fuzzy sets proved to be useful for their representation, as spatial fuzzy sets (i.e. fuzzy sets defined in the space S). Definitions of relationships have then to be extended to be applicable on fuzzy objects (the interest of fuzzy approaches for representing spatial constraints has been emphasized e.g. in [22]). Results can also be semi-quantitative, and provided in the form of intervals or fuzzy numbers. Some metric relationships, like relative directional position, belong to the second class. Even for crisp objects, fuzzy definitions are then appropriate.

3.1   Set Relationships

Computing set relationships, like inclusion, intersection, etc. if the objects are precisely defined does not call for specific developments. If the objects are imprecise, stating if they intersect or not, or if one is included in the other, becomes a matter of degree.

The degree of intersection can be defined using a supremum of a t-norm (as for fuzzy dilation):

$$\mu_{int}(\mu, \nu) = \sup_{x \in S} t[\mu(x), \nu(x)], \tag{15}$$

or using the fuzzy volume of the t-norm in order to take more spatial information into account:

$$\mu_{int}(\mu, \nu) = \frac{V_n[t(\mu, \nu)]}{\min[V_n(\mu), V_n(\nu)]}, \tag{16}$$

where

$$V_n(\mu) = \sum_{x \in S} \mu(x). \tag{17}$$

The degree of non-intersection is then simply defined by $\mu_{\neg int} = 1 - \mu_{int}$. It is interesting to note that the degree of intersection defined from a t-norm corresponds to the dilation of μ by ν at origin.

In a similar way, the degree of inclusion of ν in μ can be defined as:

$$\inf_{x \in S} T[c(\nu(x)), \mu(x)], \tag{18}$$

and corresponds to the erosion of μ by ν at origin. These morphological interpretations allow to include set relationships in the same framework as the other relations that will be detailed below.

3.2   Adjacency

Adjacency has a large interest in image processing and pattern recognition, since it denotes an important relationship between image objects or regions [48], widely used as a feature in model-based pattern recognition.

Crisp Discrete Case. Here, we restrict ourselves to the discrete case, and use discrete topology as derived from discrete connectivity for defining adjacency between two image regions X and Y, subsets of the discrete space. Let us consider an n-dimensional discrete space (typically $\mathbb{Z}^n$), and any discrete connectivity defined on this space, called c-connectivity (for instance, for $n = 3$, we may consider 6-, 18- or 26-connectivity on a cubic grid). Crisp adjacency is defined as follows, where $n_c(x, y)$ denotes the Boolean variable stating that x and y are neighbors in the sense of the discrete c-connectivity:

For any two subsets X and Y in $\mathbb{Z}^n$, X and Y are adjacent according to the c-connectivity if: $X \cap Y = \emptyset$ and $\exists x \in X, \exists y \in Y : n_c(x, y)$.

This definition can also be expressed equivalently in terms of morphological dilation, as: $X \cap Y = \emptyset$ and $D_B(X) \cap Y \neq \emptyset$, $D_B(Y) \cap X \neq \emptyset$, where B denotes the elementary structuring element associated to the c-connectivity.

Extending Adjacency to Fuzzy Objects Using Morphological Operators. A crisp definition of adjacency between crisp objects often leads to a low robustness in case of noise or segmentation errors. Let us consider for instance a problem of model-based pattern recognition, where spatial relationships are an important part of the recognition process. If two model objects are adjacent, we expect the corresponding image objects to be adjacent too, otherwise they will be difficult to recognize. However, if classical crisp adjacency is used, the fact that two objects are adjacent or not may depend on one point only and is highly prone to noise or errors in the segmentation.

In order to include possible errors or imprecision in the processing and in the recognition, the framework of fuzzy sets is very useful. Two completely different ways for representing imprecision can be considered [11]. In the first one, the satisfaction of the adjacency property between two objects is considered to be a matter of degree; this can be more appropriate than a binary index [46,47]. The second one consists in introducing imprecision in the objects themselves, and to deal with fuzzy objects. For instance, spatial imprecision due to the limited quality of image information can be represented in an adequate way by considering fuzzy objects. Then obviously adjacency is also a matter of degree. Only the second way is addressed here.

Only a few attempts in the literature address the problem of fuzzy adjacency. Fuzzy topology was introduced in [46], where a fuzzy connectivity between points is defined but without reference to the notion of fuzzy neighborhood, or to fuzzy adjacency. Similar approaches can also be found in

[47,53], where degrees of connectivity in a fuzzy set are also introduced, but neither the connectivity nor the adjacency between two fuzzy sets are defined. Rosenfeld and Klette [49] define a degree of adjacency between two crisp sets, using a geometrical approach based on the notion of "visibility" of a set from another one. This definition is then extended to degree of adjacency between two fuzzy sets. However, this definition is not symmetrical, and probably not easy to transpose to higher dimensions.

The fuzzy extension of adjacency can be obtained either by considering the constraint on the neighborhood, or by considering the morphological expression. The second possibility is presented here.

Adjacency between fuzzy sets can be defined by translating the property involving dilation in the crisp case into fuzzy terms, by using fuzzy dilation (Section 2.2). The binary concept becomes then a degree of adjacency. The degree of adjacency between μ and ν involving fuzzy dilation is then defined as:

$$\mu_{adj}(\mu, \nu) = t[\mu_{\neg int}(\mu, \nu), \mu_{int}[D_B(\mu), \nu], \mu_{int}[D_B(\nu), \mu]]. \tag{19}$$

This definition represents a conjunctive combination of a degree of non-intersection $\mu_{\neg int}$ between μ and ν and a degree of intersection μ_{int} between one fuzzy set and the dilation of the other. B can be taken as the elementary structuring element related to the considered connectivity, or as a fuzzy structuring element, representing for instance spatial imprecision (i.e. the possibility distribution of the location of each point). We proved that this definition is symmetrical, consistent with the binary definition if μ, ν and B are binary, decreases if the distance between μ and ν increases, and is invariant with respect to geometrical transformations [11].

Since in the discrete binary case the equation using dilation means that the minimum (nearest point) distance between X and Y is equal to 1, we can also exploit this fact in the fuzzy case, by using the fuzzy nearest point distance, defined from fuzzy dilation (see Section 3.3). It leads to similar definitions, sharing the same properties as the previous ones:

$$\mu_{adj}(\mu, \nu) = \delta_N(\mu, \nu)(1), \tag{20}$$

where $\delta_N(\mu, \nu)(1)$ is obtained as in Equation 30, by taking $n = 1$.

3.3   Distances

The importance of distances in image processing is well established. Their extensions to fuzzy sets can be useful in several parts of image processing under imprecision. Let us mention a few possible applications of these distances for image processing problems where imprecision has to be taken into account. Distance from a point to a fuzzy set can be used for classification purposes, where a point has to be attributed to the nearest fuzzy class. When considering distance from a point to the complement of a fuzzy set μ, we obtain the basic information for computing a fuzzy skeleton of μ. Mean distance is useful for registration: if we want to register a fuzzy set with respect to another one, we may use this distance as a minimization criterion, that can be optimized over all possible positions (typically translation and rotation) of the one fuzzy set with respect to the other.

Several definitions can be found in the literature for distances between fuzzy sets (which is the main addressed problem). They can be roughly divided in two classes: distances that take only membership functions into account and that compare them pointwise, and distances that additionally include spatial distances. The definitions which combine spatial distance and fuzzy membership comparison allow for a more general analysis of structures in images, for applications where topological and spatial arrangement of the structures of interest is important (segmentation, classification, scene interpretation). This is permitted by the fact that these distances combine membership values at different points in the space, and take into account their proximity or distance in S. The price to pay is an increased complexity, generally quadratic in the cardinality of S.

We proposed in [8] original approaches for defining fuzzy distances taking into account spatial information, which are based on fuzzy mathematical morphology. They are summarized below. The idea is that in the binary case, there exist strong links between mathematical morphology (in particular dilation) and distances (from a point to a set, and between two sets), and this can also be exploited in the fuzzy case. The advantage is that distances are expressed in set theoretical terms, and are therefore easier to translate with nice properties than usual analytical expressions.

Distances from a Point to a Set or a Fuzzy Set. A formalism where distances are expressed in set theoretical terms is provided by mathematical morphology, since the distance from a point to a set can be expressed in terms of morphological dilation. In the crisp case, and in a finite discrete space (which is the most interesting case in image processing), we have respectively for $n = 0$ and for $n > 0$:

$$d(x, X) = 0 \Leftrightarrow x \in X \qquad (21)$$
$$d(x, X) = n \Leftrightarrow x \in D^n(X) \text{ and } x \notin D^{n-1}(X) \qquad (22)$$

where D^n denotes the dilation by a ball of radius n centered at the origin of S (and $D^0(X) = X$) (see e.g. [13] for a study of discrete balls and discrete distances in the crisp case). In this case, the extensivity property of the dilation holds [50], and $x \notin D^{n-1}(X)$ is equivalent to $\forall n' < n, x \notin D^{n'}(X)$. Equation 22 is equivalent to: $x \in D^n(X) \cap [D^{n-1}(X)]^C$, where A^C denotes the complement set of A in S.

This is a pure set theoretical expression, that we can now translate into fuzzy terms in order to define the distance from a point x to a spatial fuzzy set μ [8]. This leads to the following definition of the degree to which $d(x, \mu)$

is equal to n:

$$\delta_{(x,\mu)}(0) = \mu(x), \tag{23}$$

$$\delta_{(x,\mu)}(n) = t[D_\nu^n(\mu)(x), c[D_\nu^{n-1}(\mu)(x)]], \tag{24}$$

where t is a t-norm (fuzzy intersection), c a fuzzy complementation (typically $c(a) = 1 - a$ for $a \in [0,1]$), and ν a fuzzy structuring element used for performing the dilation. Several choices of ν are possible. It can be simply the unit ball, or a fuzzy set representing for instance the smallest sensitive unit in the image, along with the imprecision attached to it. In this case, ν has to be equal to 1 at the origin of S, such that the extensivity of the dilation still holds [10].

The properties of this definition are the following: If $\mu(x) = 1$, $\delta_{(x,\mu)}(0) = 1$ and $\forall n > 0, \delta_{(x,\mu)}(n) = 0$, i.e. the distance is a crisp number in this case. If μ and ν are binary, the proposed definition coincides with the binary one. The fuzzy set $\delta_{(x,\mu)}$ can be interpreted as a density distance, from which a distance distribution can be deduced by integration. Finally, $\delta_{(x,\mu)}$ is a non normalized fuzzy number (in the discrete finite case).

Distances Between Two Sets or Two Fuzzy Sets. Distances between two objects can be expressed in morphological terms and then extended to distances between fuzzy sets [8]. Thanks to the algebraic framework provided by morphological expressions, this extension is possible by a direct translation of crisp equations into fuzzy ones (and this is easier than translating usual analytical expressions of distances). We just give the examples of nearest point distance and Hausdorff distance.

Fuzzy nearest point distance. The minimum or nearest point distance between X and Y is defined (in the discrete finite case) as:

$$d_N(X,Y) = \min_{(x,y)\in X\times Y} d_E(x,y) = \min_{x\in X} d_E(x,Y) = \min_{y\in Y} d_E(y,X), \tag{25}$$

where d_E denotes the Euclidean distance in S. This has an equivalent morphological expression:

$$d_N(X,Y) = \inf\{n \in \mathbb{N}, X \cap D^n(Y) \neq \emptyset\} = \inf\{n \in \mathbb{N}, Y \cap D^n(X) \neq \emptyset\}. \tag{26}$$

By translating equation 26, we define a distance distribution $\Delta_N(\mu,\mu')(n)$ that expresses the degree to which the distance between μ and μ' is less than n by:

$$\Delta_N(\mu,\mu')(n) = f[\sup_{x\in S} t[\mu(x), D_\nu^n(\mu')(x)], \sup_{x\in S} t[\mu'(x), D_\nu^n(\mu)(x)]], \tag{27}$$

where f is a symmetrical function.

A distance density, i.e. a fuzzy number $\delta_N(\mu, \mu')(n)$ representing the degree to which the distance between μ and μ' is equal to n, can be obtained implicitly by $\Delta_N(\mu, \mu')(n) = \int_0^n \delta_N(\mu, \mu')(n')dn'$. Clearly, this expression is not very tractable and does not lead to a simple explicit expression of $\delta_N(\mu, \mu')(n)$. Therefore, we suggest to use an explicit method, exploiting the fact that, for $n > 0$, the nearest point distance can be expressed in morphological terms as:

$$d_N(X, Y) = n \Leftrightarrow D^n(X) \cap Y \neq \emptyset \text{ and } D^{n-1}(X) \cap Y = \emptyset \qquad (28)$$

or equivalently by the symmetrical expression. For $n = 0$ we have:

$$d_N(X, Y) = 0 \Leftrightarrow X \cap Y \neq \emptyset. \qquad (29)$$

The translation of these equivalences provides, for $n > 0$, the following distance density:

$$\delta_N(\mu, \mu')(n) = t[\sup_{x \in S} t[\mu'(x), D_\nu^n(\mu)(x)], c[\sup_{x \in S} t[\mu'(x), D_\nu^{n-1}(\mu)(x)]]] \qquad (30)$$

or a symmetrical expression derived from this one, and:

$$\delta_N(\mu, \mu')(0) = \sup_{x \in S} t[\mu(x), \mu'(x)]. \qquad (31)$$

Fuzzy Hausdorff distance. Like for the nearest point distance, we can extend the Hausdorff distance by translating directly the binary equation defining the Hausdorff distance:

$$d_H(X, Y) = \max[\sup_{x \in X} d(x, Y), \sup_{y \in Y} d(y, X)]. \qquad (32)$$

This distance can be expressed in morphological terms as:

$$d_H(X, Y) = \inf\{n, X \subset D^n(Y) \text{ and } Y \subset D^n(X)\}. \qquad (33)$$

From equation 33, a distance distribution can be defined, by introducing fuzzy dilation:

$$\Delta_H(\mu, \mu')(n) = t[\inf_{x \in S} T[D_\nu^n(\mu)(x), c(\mu'(x))], \inf_{x \in S} T[D_\nu^n(\mu')(x), c(\mu(x))]], \qquad (34)$$

where c is a complementation, t a t-norm and T a t-conorm. A distance density can be derived implicitly from this distance distribution.

A direct definition of a distance density can be obtained from:

$$d_H(X, Y) = 0 \Leftrightarrow X = Y, \qquad (35)$$

and for $n > 0$:

$$d_H(X, Y) = n \Leftrightarrow X \subset D^n(Y) \text{ and } Y \subset D^n(X)$$
$$\text{and } \left(X \not\subset D^{n-1}(Y) \text{ or } Y \not\subset D^{n-1}(X)\right). \qquad (36)$$

Translating these equations leads to a definition of the Hausdorff distance between two fuzzy sets μ and μ' as a fuzzy number:

$$\delta_H(\mu, \mu')(0) = t[\inf_{x \in S} T[\mu(x), c(\mu'(x))], \inf_{x \in S} T[\mu'(x), c(\mu(x))]], \qquad (37)$$

$$\delta_H(\mu, \mu')(n) = t[\inf_{x \in S} T[D_\nu^n(\mu)(x), c(\mu'(x))], \inf_{x \in S} T[D_\nu^n(\mu')(x), c(\mu(x))],$$

$$T(\sup_{x \in S} t[\mu(x), c(D_\nu^{n-1}(\mu')(x))], \sup_{x \in S} t[\mu'(x), c(D_\nu^{n-1}(\mu)(x))])]. \qquad (38)$$

Properties. The above definitions of fuzzy nearest point and Hausdorff distances (defined as fuzzy numbers) between two fuzzy sets do not necessarily share the same properties as their crisp equivalent. This is due in particular to the fact that, depending on the choice of the involved t-norms and t-conorms, excluded-middle and non-contradiction laws may not be satisfied. All distances are positive, in the sense that the defined fuzzy numbers have always a support included in $\mathbb{R}^+$. By construction, all defined distances are symmetrical with respect to μ and μ'. The separability property (i.e. $d(\mu, \nu) = 0 \Leftrightarrow \mu = \nu$) is not always satisfied. However, if μ is normalized, we have for the nearest point distance $\delta_N(\mu, \mu)(0) = 1$ and $\delta_N(\mu, \mu)(n) = 0$ for $n > 1$. For the Hausdorff distance, $\delta_H(\mu, \mu')(0) = 1$ implies $\mu = \mu'$ for T being the bounded sum $(T(a, b) = \min(1, a + b))$, while it implies μ and μ' crisp and equal for $T = \max$. Also the triangular inequality is not satisfied in general.

3.4   Directional Relative Position from Conditional Fuzzy Dilation

Relationships between objects can be partly described in terms of relative position, like "to the left of". It should be noted that such concepts are rather ambiguous, they defy precise definitions, but human beings have a rather intuitive and common way of understanding and interpreting them. From our every day experience, it is clear that any "all-or-nothing" definition leads to unsatisfactory results in several situations, even of moderate complexity. Therefore, relative position concepts may find a better understanding in the framework of fuzzy sets, as fuzzy relationships, even for crisp objects. This framework makes it possible to propose flexible definitions which fit the intuition and may include subjective aspects, depending on the application and on the requirements of the user.

Overview: Morphological Fuzzy Pattern Matching Approach. Our motivation for proposing a definition for relative position between objects is to provide a definition that should: work directly in the image space; be generic enough in order to apply to relative positions defined by any direction (not only four or six basic ones); introduce morphological information on the considered objects themselves, with the aim of pattern recognition applications; be applicable to 3D objects (for applications to medical imaging for instance) and to fuzzy objects (in order to take into account spatial imprecision in the objects); verify algebraic and geometrical properties and behave according to the intuition in a large variety of situations.

Let us consider a reference object R and an object A for which the relative position with respect to R has to be evaluated. In order to evaluate the degree to which A is in some direction with respect to R, we propose the following approach [6,7]:

1. We first define a fuzzy "landscape" around the reference object R as a fuzzy set such that the membership value of each point corresponds to the degree of satisfaction of the spatial relation under examination. Therefore the fuzzy landscape is directly defined in the same space as the considered objects, in the contrary to the solution proposed in [31] (this idea will be further exploited in Section 4). This also contrasts with methods based on centroids [32], on angle histograms [40,30], or on force histograms [39].

2. We then compare the object A to the fuzzy landscape attached to R, in order to evaluate how well the object matches with the areas having high membership values (i.e. areas that are in the desired direction). This is done using a fuzzy pattern matching approach, which provides an evaluation as an interval instead of one number only. This makes a major difference with respect to all the previous approaches, and, to our opinion, it provides a richer information about the considered relationship.

Relative Position from Fuzzy Pattern Matching. In the 3D Euclidean space $\mathcal{S}$, a direction is defined by two angles α_1 and α_2, where $\alpha_1 \in [0, 2\pi]$ and $\alpha_2 \in [-\frac{\pi}{2}, \frac{\pi}{2}]$ ($\alpha_2 = 0$ in the 2D case). The direction in which the relative position of an object with respect to another one is evaluated is denoted by: $u_{\alpha_1, \alpha_2} = (\cos \alpha_2 \cos \alpha_1, \cos \alpha_2 \sin \alpha_1, \sin \alpha_2)^t$, and we note $\alpha = (\alpha_1, \alpha_2)$.

We consider two (possibly fuzzy) objects, R and A, and define the degree to which A is in direction u_{α_1, α_2} with respect to R. Let us denote by $\mu_\alpha(R)$ the fuzzy set defined in the image in such a way that points of areas which satisfy to a high degree the relation "to be in the direction u_{α_1, α_2} with respect to reference object R" have high membership values. In other terms, the membership function $\mu_\alpha(R)$ has to be an increasing function of the degree of satisfaction of the relation. It is a spatial fuzzy set (i.e. a function of the image space $\mathcal{S}$ into $[0,1]$) and directly related to the shape of R. The precise definition of $\mu_\alpha(R)$ is given in Section 4.5, and is based on fuzzy dilation.

Let us denote by μ_A the membership function of the object A, which is a function of $\mathcal{S}$ into $[0,1]$. The evaluation of relative position of A with respect to R is given by a function of $\mu_\alpha(R)(x)$ and $\mu_A(x)$ for all x in $\mathcal{S}$. An appropriate tool for defining this function is the fuzzy pattern matching approach [21]. Following this approach, the evaluation of the matching between two possibility distributions consists of two numbers, a necessity degree N (a pessimistic evaluation) and a possibility degree Π (an optimistic evaluation), as often used in the fuzzy set community. In our application, they take the following forms:

$$\Pi^R_{\alpha_1,\alpha_2}(A) = \sup_{x \in S} t[\mu_\alpha(R)(x), \mu_A(x)], \tag{39}$$

$$N^R_{\alpha_1,\alpha_2}(A) = \inf_{x \in S} T[\mu_\alpha(R)(x), 1 - \mu_A(x)], \tag{40}$$

where t is a t-norm (fuzzy intersection) and T a t-conorm (fuzzy union) [20]. In the crisp case, these equations reduce to: $\Pi^R_{\alpha_1,\alpha_2}(A) = \sup_{x \in A} \mu_\alpha(R)(x)$, and $N^R_{\alpha_1,\alpha_2}(A) = \inf_{x \in A} \mu_\alpha(R)(x)$.

The possibility corresponds to a degree of intersection between the fuzzy sets A and $\mu_\alpha(R)$, while the necessity corresponds to a degree of inclusion of A in $\mu_\alpha(R)$. They can also be interpreted in terms of fuzzy mathematical morphology, since the possibility $\Pi^R_{\alpha_1,\alpha_2}(A)$ is equal to the dilation of μ_A by $\mu_\alpha(R)$ at the origin, while the necessity $N^R_{\alpha_1,\alpha_2}(A)$ is equal to the erosion, as shown in [10]. These two interpretations, in terms of set theoretic operations and in terms of morphological ones, explain how the shape of the objects is taken into account.

Several other functions combining $\mu_\alpha(R)$ and $\mu_A(x)$ can be constructed. The extreme values provided by the fuzzy pattern matching are interesting because of their morphological interpretation, and because they provide an interval and not only a single value and may represent in this way the ambiguity of the relation if any. An average measure can also be useful from a practical point of view, and is defined as:

$$M^R_{\alpha_1,\alpha_2}(A) = \frac{1}{|A|} \sum_{x \in S} \mu_A(x)\mu_\alpha(R)(x), \tag{41}$$

where $|A|$ denotes the fuzzy cardinality of A: $|A| = \sum_{x \in S} \mu_A(x)$.

Properties. We proved [7] that the possibility has a symmetry property (i.e. the possibility for A to be in some direction from B is equal to the possibility of B to be in the opposite direction with respect to A). Also, the proposed definition is invariant with respect to translation, rotation and scaling, for 2D and 3D objects (crisp and fuzzy). We have also studied the influence of the distance between objects on their relative position. We proved that when the distance between the objects increases, the objects are seen as points. The value of their relative position can be predicted only from the direction of interest and the direction in which one object goes far away from the reference object. Therefore the shape of the objects does no longer play any role in the assessment of their relative position. Finally, we looked at the behavior of the proposed definition on cases where the reference object has strong concavities, and show that the behavior corresponds to what can be intuitively expected.

3.5 Example

As an illustrative example, we extracted from a brain MRI image a few internal structures which are represented in Figure 1 as spatial fuzzy sets, where membership degrees are represented using grey levels. The use of fuzzy sets may represent different types of imprecision, either on the boundary of the objects (due for instance to partial volume effect, or to the spatial resolution), or on the individual variability of these structures, etc.

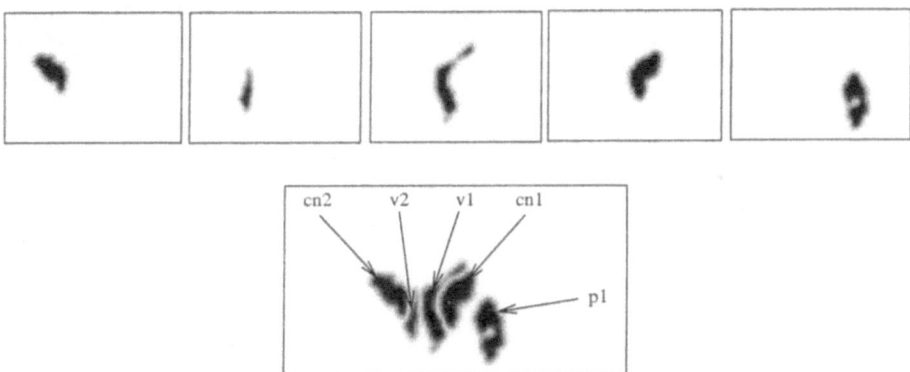

Fig. 1. Top: 5 fuzzy objects representing internal brain structures (top part of the ventricular system and surrounding objects) segmented in a MRI brain image (membership values rank between 0 and 1, from white to black). From left to right: right caudate nucleus (cn2), right lateral ventricle (v2), left lateral ventricle (v1), left caudate nucleus (cn1), left putamen (p1) (using the standard "left is right" convention, e.g. cn1 is on the right in this figure). Bottom: superposition of these fuzzy objects (the maximum membership value is displayed at each point).

The adjacency degrees between some of these fuzzy objects are given in Table 1. The results are in agreement with what can be expected from the model (crisp adjacency between atlas objects). In this case, crisp adjacency would provide completely different results in the model and in the image, preventing its use for recognition. This suggests that fuzzy adjacency degree can indeed be used for pattern recognition purposes, of course combined with other spatial relationships.

Examples of distances between brain structures are shown in Table 2, as fuzzy numbers issued from the morphological definition of Hausdorff distance. The results are in agreement with what is expected: the model of v2 provided by an anatomical atlas is near from cn2 and v1, quite far from cn1 and very far from p1. We do not obtain a null value for v2, since it does not perfectly match the model of v2, but we obtain values that are still much lower than those obtained for the other structures. This shows that distances can be used both for identifying a structure among several possible ones, by using distance as a dissimilarity measure, and for describing the spatial arrangement of objects.

Fuzzy object 1	Fuzzy object 2	degree of adjacency	adjacency in the model (crisp)
v1	v2	0.368	1
v1	cn1	0.463	1
v1	p1	0.000	0
v1	cn2	0.035	0
v2	cn2	0.427	1
cn1	p1	0.035	0

Table 1. Results obtained for fuzzy adjacency. Labels of structures are given in Figure 1. High degrees are obtained between structures where adjacency is expected, while very low degrees are obtained in the opposite case.

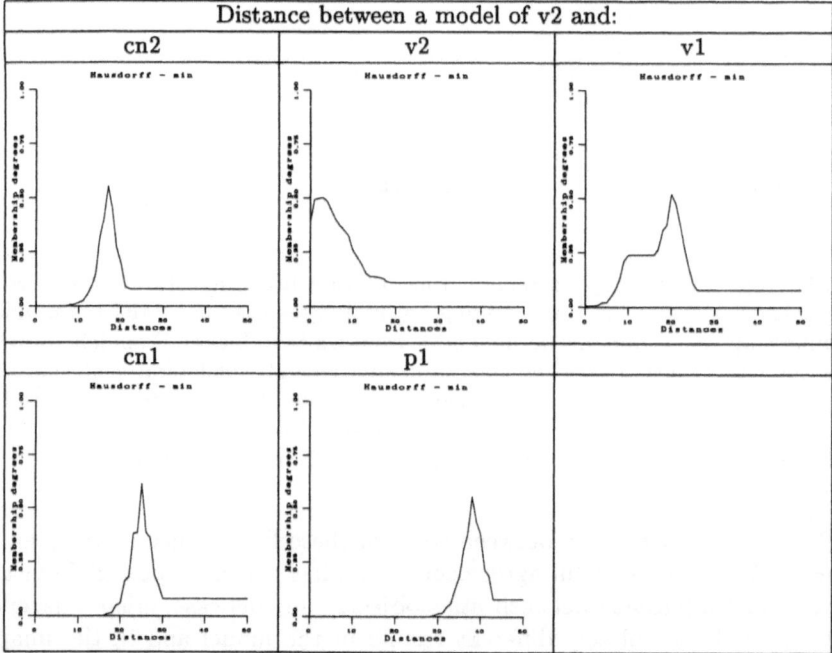

Table 2. Distances between fuzzy sets using the morphological approach, for the Hausdorff distance, using minimum as t-norm. The distance is computed between each of the 5 structures and a model of v2 given by an anatomical atlas.

The fuzzy landscapes representing the degree of satisfaction of the relations "left to", "right to", " below" and "above" object v1 are shown in Figure 2. The relative position degrees between some of the obtained fuzzy objects are given in Figure 3. The interpretation of these results is straightforward with respect to the intuitive expected relative positions. Object cn1 is mainly to the right of v1 (and only with very low degree to its left), and

quite above and below. This expresses that it is "in the right concavity of
v1", an example of more complex relationship derived from the basic relative
positions. Object cn2 is to the left of v1, with no ambiguity at all concerning
the right relationship (i.e. no point of cn2 is to the right of v1). It is quite
above v1, and less below it than cn1. Similar interpretations can be given for
p1 and v2 with respect to v1.

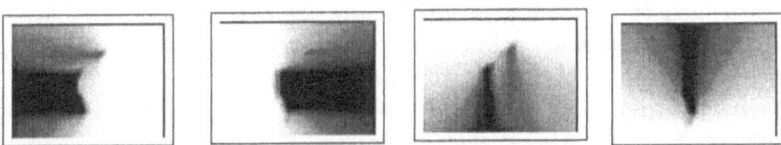

Fig. 2. Fuzzy areas corresponding to four relationships of Figure 3 for the object
v1 of Figure 1.

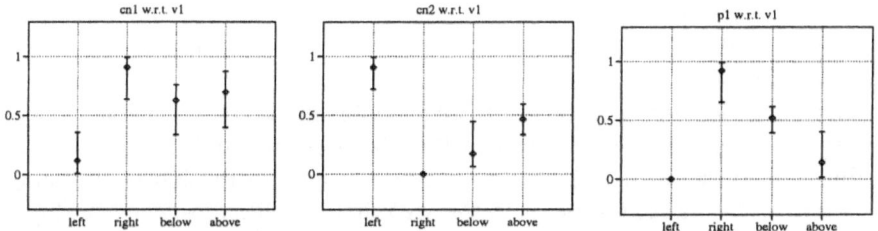

Fig. 3. Results of relative position obtained for some of the objects of Figure 1.
Vertical bars represent $[N, \Pi]$ intervals and diamonds the average values.

The use of this results for recognizing brain structures from an atlas calls
for defining a morphism between the image objects and the atlas involving
similarities between spatial relations in the model and in the image [41] and
optimizing it [3,42].

4   Spatial Representations of Spatial Relationships

4.1   Spatial Fuzzy Sets as a Representation Framework

Usually vision and image processing make use of quantitative representations
of spatial relationships. In artificial intelligence, mainly symbolic representations are developed (see [55] for a survey). Limitations of purely qualitative reasoning has already been stressed in [22], as well as the interest of
adding semiquantitative extension to qualitative value (as done in the fuzzy
set theory for linguistic variables [57,19]) for deriving useful and practical conclusions (as for recognition). Purely quantitative representations are limited

in the case of imprecise statements, and of knowledge expressed in linguistic terms. Here we propose to integrate both quantitative and qualitative knowledge, using semiquantitative interpretation of fuzzy sets. As already mentioned in [23], this allows to provide a computational representation and interpretation of imprecise spatial constraints, expressed in a linguistic way, possibly including quantitative knowledge.

Knowledge used for guiding recognition of an object or to perform spatial reasoning is generally heterogeneous. It may concern the object we are looking at (its shape, topology, color, position), or relationships to other objects (distances, adjacency, relative directional position). It may be generic (typically if derived from a model or from expert knowledge), or factual (if derived from the scene itself). And it may be usually provided in a lot of different forms. Classically it can be a number, a distribution or a binary value. But we will also be concerned with imprecise values and with propositional formulas which are often used by experts within a given application. Imprecise values are expressed sometimes in linguistic terms: for instance the expected distance between two objects (*close, far, ...*). They can also be expressed as an interval. Propositional formulas (*object A is to the right of object B*) usually express rather complex ideas which need much prior knowledge to be correctly interpreted.

The main idea here is to translate all available knowledge as a spatial representation. Such representations can then be used in a fusion process that combines all these regions of interest in order to focus attention by reducing the search space and to restrict it to the area that satisfies most relationships. Since many pieces of information are delivered in an imprecise way, we make use of the framework of fuzzy sets [56,19]. This modeling is well adapted to information derived from a schematic model and from images. This framework provides a good theoretical basis to model the imprecision of the information at different levels of representation, and to represent both numerical and symbolic information, including structural information (constituted mainly by spatial relationships here). An illustration of this for structural model-based pattern recognition is described in [27].

In the following a point (volume element or voxel) in the 3D discrete space S is denoted by v. For each piece of knowledge, we consider its "natural expression", i.e. the usual form in which it is given or available, and translate it into a spatial fuzzy set in the space, the membership of which is denoted by $\mu_{knowledge}$:

$$\mu_{knowledge} : \begin{cases} S \to [0,1] \\ v \mapsto \mu_{knowledge}(v). \end{cases} \qquad (42)$$

In this representation, each piece of knowledge becomes a fuzzy region of the space. If the knowledge is considered as a constraint to be satisfied by the object to be recognized, this fuzzy region represents a search area or a fuzzy volume of interest for this object. This type of representation provides a common framework to represent pieces of information of various types (objects, spatial imprecision, relationships to other objects, etc.). Therefore the fuzzy regions defined in the space $\mathcal{S}$ corresponding to these pieces of information may have different semantics. Moreover, this common framework allows the combination of this heterogeneous information.

The numerical representation of membership values assumes that we can assign numbers that represent degrees of satisfaction of a relationship for instance. These numbers can be derived from prior knowledge or learned from examples, but usually there remain some quite arbitrary choices. This might appear as a drawback in comparison to propositional representations. However, it is not necessary to have precise estimations of these values, and experimentally we observed a good robustness with respect to these estimations, in various problems like information fusion, object recognition and scene interpretation [12,27]. This can be explained by two reasons: first, the fuzzy representations are used for rough information and therefore do not have to be precise itself, and second several pieces of information are usually combined in a whole reasoning process, which decreases the influence of each particular value (of individual information). Therefore the chosen numbers are not crucial. What is important is that ranking is preserved. For instance if a region of the space satisfies a relationship to some objects to a higher degree that another region, then this ranking is preserved in the representation, for all relationships described in the following sections. Assuming the existence of ranking is reasonable for the type of relations we consider.

4.2 Set Relationships

Set relationships specify if areas where other objects can be localized are forbidden or obligatory. This goes with the idea of progressively exploring the space. These set relationships are expressed as inclusion in objects (denoted by O^{in}) or exclusion from objects (denoted by O^{out}). The corresponding region of interest has the following membership function:

$$\forall v \in \mathcal{S}, \ \mu_{\text{set}}(v) = \begin{cases} 1 & \text{if } v \in O^{\text{in}} \setminus O^{\text{out}} \\ 0 & \text{elsewhere.} \end{cases} \tag{43}$$

If O^{in} and O^{out} are fuzzy objects, defined on $\mathcal{S}$ through their membership functions $\mu_{O^{\text{in}}}$ and $\mu_{O^{\text{out}}}$, then the previous equation becomes:

$$\mu_{\text{set}}(v) = t[\mu_{O^{\text{in}}}(v), 1 - \mu_{O^{\text{out}}}(v)], \tag{44}$$

where t is a t-norm, which expresses a conjunction between inclusion constraint and exclusion constraint. The properties of t-norms guarantee that intuitive requirements are satisfied:

- Since any t-norm is smaller than the min, μ_{set} is included in $\mu_{O^{\text{in}}}$.
- Similarly, μ_{set} is included in the complement of $\mu_{O^{\text{out}}}$.

- Increasingness of t induces increasingness of μ_{set} with respect to $\mu_{O^{in}}$ and decreasingness with respect to $\mu_{O^{out}}$, which expresses that constraining a fuzzy region to be included in smaller objects (respectively excluded from larger objects) leads to a smaller fuzzy region.
- Commutativity and associativity of t-norms allow to introduce constraints on inclusion in several objects (or exclusion from several objects) in any order.

4.3   Adjacency

Here we extend previous work [11] for defining fuzzy regions expressing an adjacency constraint with an object. Fuzzy dilation can be used to define the external boundary of a fuzzy object μ, as:

$$b_e(\mu) = t[D_\nu(\mu), 1 - \mu]. \tag{45}$$

In a similar way, internal boundary is defined from fuzzy erosion, as:

$$b_i(\mu) = t[\mu, D_\nu(1 - \mu)] = t[\mu, 1 - E_\nu(\mu)], \tag{46}$$

where $E_\nu(\mu)$ denotes the fuzzy erosion of μ by ν [10]:

$$\forall v \in \mathcal{S}, \; E_\nu(\mu)(v) = \inf_{v' \in \mathcal{S}} T[\mu(v'), 1 - \nu(v' - v)], \tag{47}$$

where T is a t-conorm.

The search area for an object adjacent to μ is then expressed by two fuzzy sets: $b_e(\mu)$ that should be intersected by the boundary of the searched object, and $1 - \mu$ which expresses the inclusion constraint in the complement of μ.

The link between adjacency and nearest distance (see Section 3.2) could also be used here.

From these basic topological relationships (inclusion, exclusion, adjacency), other ones can be derived. For instance, an object that is a tangential proper part of μ has to be searched in μ and its boundary has to intersect $b_i(\mu)$. A non tangential proper part of μ has to be searched in $E_\nu(\mu)$.

4.4   Distances

In the framework of our study, expressions of knowledge about distances will be translated in spatial volumes of interest within $\mathcal{S}$, taking into account imprecision and uncertainty, in order to account for approximate statements where distances can be expressed as numbers, but also intervals, fuzzy numbers, linguistic values, etc.

In contrary to the approach proposed in [25,26] where linguistic variables about distances are represented as fuzzy sets on each axis, from which distance knowledge in the space can be derived, we choose here to represent distance knowledge directly in the space $\mathcal{S}$, as spatial fuzzy sets. The method we propose is independent of the dimension of $\mathcal{S}$ and uses morphological expressions of distances [8], as detailed in Section 3.3.

We assume that a set A is known as one already recognized object, or a known area of $\mathcal{S}$, and that we want to determine B, subject to satisfy some distance relationship with A. According to the algebraic expressions of distances, dilation of A is an adequate tool for this. Let us consider the following different cases:

- If knowledge expresses that $d(A, B) = n$, then the border of B should intersect the region defined by $D^n(A) \setminus D^{n-1}(A)$, which is made up of the points exactly at distance n from A, and B should be looked for in $D^{n-1}(A)^C$ (the complement of the dilation of size $n - 1$).
- If knowledge expresses that $d(A, B) \leq n$, then B should be looked for in A^C, with the constraint that at least one point of B belongs to $D^n(A) \setminus A$.
- If knowledge expresses that $d(A, B) \geq n$, then B should be looked for in $D^{n-1}(A)^C$.
- If knowledge expresses that $n_1 \leq d(A, B) \leq n_2$, then B should be searched in $D^{n_1-1}(A)^C$ with the constraint that at least one point of B belongs to $D^{n_2}(A) \setminus D^{n_1-1}(A)$.

The constraints on the border lead to the definition of actually two fuzzy sets, one for constraining the object, and one constraining its border, as for adjacency. However, they can be avoided by considering both minimum and maximum (Hausdorff) distances, expressing for instance that B should lay between a distance n_1 and a distance n_2 of A. Therefore, the minimum distance should be greater than n_1 and the maximum distance should be less than n_2. In this case, the volume of interest for B is reduced to $D^{n_2}(A) \setminus D^{n_1-1}(A)$.

In cases where imprecision has to be taken into account, fuzzy dilations are used, with the corresponding equivalences with fuzzy distances [10,8]. The extension to approximate distances calls for fuzzy structuring elements. We define these structuring elements through their membership function ν on $\mathcal{S}$. Structuring elements with a spherical symmetry can typically be used, where the membership degree only depends on the distance to the center of the structuring element.

Let us consider the generalization to the fuzzy case of the last case (minimum distance of at least n_1 and maximum distance of at most n_2 to a fuzzy set μ). Instead of defining an interval $[n_1, n_2]$, we consider a fuzzy interval, defined as a fuzzy set on $\mathbb{R}^+$ having a core equal to the interval $[n_1, n_2]$. The membership function μ_n is increasing between 0 and n_1 and decreasing after n_2 (this is but one example). Then we define two structuring elements, as:

$$\nu_1(v) = \begin{cases} 1 - \mu_n(d_E(v, 0)) & \text{if } d_E(v, 0) \leq n_1 \\ 0 & \text{else} \end{cases} \tag{48}$$

$$\nu_2(v) = \begin{cases} 1 & \text{if } d_E(v,0) \leq n_2 \\ \mu_n(d_E(v,0)) & \text{else} \end{cases} \tag{49}$$

where d_E is the Euclidean distance in S and O the origin. The spatial fuzzy set expressing the approximate relationship about distance to μ is then defined as:

$$\mu_{\text{distance}} = t[D_{\nu_2}(\mu), 1 - D_{\nu_1}(\mu)] \tag{50}$$

if $n_1 \neq 0$, and $\mu_{\text{distance}} = D_{\nu_2}(\mu)$ if $n_1 = 0$. The increasingness of fuzzy dilation with respect to both the set to be dilated and the structuring element [10] guarantees that these expressions do not lead to inconsistencies. Indeed, we have $\nu_1 \subset \nu_2$, $\nu_1(0) = \nu_2(0) = 1$, and therefore $\mu \subset D_{\nu_1}(\mu) \subset D_{\nu_2}(\mu)$. In the case where $n_1 = 0$, we do not have $\nu_1(0) = 1$ any longer, but in this case, only the dilation by ν_2 is considered. This case corresponds actually to a distance to μ less than "about n_2". These properties are indeed expected for representations of distance knowledge.

Figure 4 illustrates this approach. The two structuring elements ν_1 and ν_2 are derived from a fuzzy interval μ_n, are used for dilation of an object on the left (buildings extracted from a map), and μ_{distance} is computed to represent the approximate knowledge about the distance to this object.

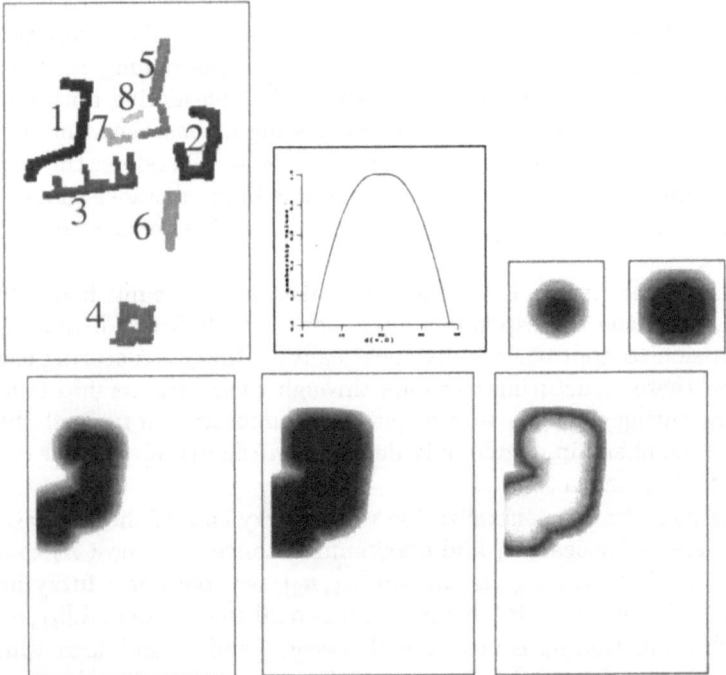

Fig. 4. Buildings extracted from a map, membership function μ_n, structuring elements ν_1 and ν_2, dilation of building 1 with these two structuring elements, and representation of μ_{distance}.

From an algorithmic point of view, fuzzy dilations may be quite heavy if the structuring element has a large support. However, in the case of crisp objects and structuring elements with spherical symmetry, fast algorithms can be implemented. The distance to the object A is first computed using chamfer algorithms [13]. It defines a distance map in $\mathcal{S}$, which gives the distance of each voxel v to object A. This discrete distance can be made as precise as necessary [36]. Then the translation into a fuzzy volume of interest is made according to a simple look-up table derived from μ_n. This algorithm has a linear complexity in the cardinality of $\mathcal{S}$.

4.5   Relative Directional Position

The definition of directional position between two sets described in Section 3.4 relies on a spatial representation of the degree of satisfaction of the relation to the reference object, which is now detailed.

We denote by $\mu_\alpha(A)$ the fuzzy region representing the relation *to be in the direction* $\boldsymbol{u}_{\alpha_1,\alpha_2}$ *with respect to reference object* A. Points that satisfy this relation with high degrees should have high membership values. In other terms, the membership function $\mu_\alpha(A)$ has to be an increasing function of the degree of satisfaction of the relation.

Let us denote by P any point in $\mathcal{S}$, and by Q any point in A. Let $\beta(P,Q)$ be the angle between the vector $\boldsymbol{QP}$ and the direction $\boldsymbol{u}_{\alpha_1,\alpha_2}$, computed in $[0, \pi]$:

$$\beta(P,Q) = \arccos\left[\frac{\boldsymbol{QP} \cdot \boldsymbol{u}_{\alpha_1,\alpha_2}}{\|\boldsymbol{QP}\|}\right], \text{ and } \beta(P,P) = 0. \tag{51}$$

Setting $\beta(P,P) = 0$ allows actually to deal with overlapping objects or with fuzzy objects with overlapping supports.

We then determine for each point P the point Q of A leading to the smallest angle β, denoted by $\beta_{\min}$. In the crisp case, this point Q is the reference object point from which P is visible in the direction the closest to $\boldsymbol{u}_{\alpha_1,\alpha_2}$ (see Figure 5): $\beta_{\min}(P) = \min_{Q \in A} \beta(P,Q)$.

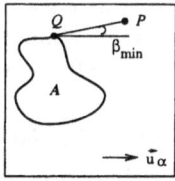

Fig. 5. Definition of $\beta_{\min}$ with respect to an object A.

The spatial fuzzy set $\mu_\alpha(A)$ at point P is then defined as:

$$\mu_\alpha(A)(P) = f(\beta_{\min}(P)), \tag{52}$$

where f is a decreasing function of $[0, \pi]$ into $[0, 1]$. In our experiments, we have chosen a simple linear function: $\mu_\alpha(A)(P) = \max(0, 1 - \frac{2\beta_{\min}(P)}{\pi})$, but other functions can be used, as trigonometric functions. We chose a function that sets the values of $\mu_\alpha(A)(P)$ to 0 as soon as $\beta_{\min}$ becomes greater than $\pi/2$. This avoids to get positive membership values for points having coordinates completely outside of the coordinate range of A in the desired direction.

In the fuzzy case, we propose a method which translates binary equations and propositions into fuzzy ones: in the binary case, we express that: $Q \in A$ and $f(\beta_{\min}) = \max_{Q \in A} f(\beta(P, Q))$ (since f is decreasing), which translates in fuzzy terms as:

$$\mu_\alpha(A)(P) = \max_{Q \in Supp(A)} t[\mu_A(Q), f(\beta(P, Q))], \qquad (53)$$

where t is a t-norm.

Figure 6 illustrates a result obtained with this method, on one object of Figure 4. It shows that it well fits the intuition.

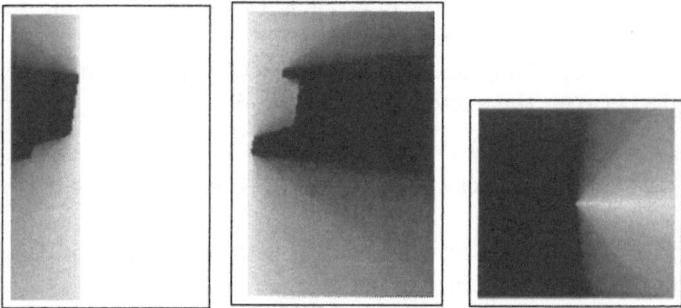

Fig. 6. Left: fuzzy region representing the relationship "to the left of building 1" using Equation 53. Middle: fuzzy region representing the relationship "to the right of building 1". Right: Structuring element ν for $\alpha = 0$.

An advantage of this approach is its easy interpretation in terms of morphological operations. It can indeed be shown [6] that $\mu_\alpha(A)$ is exactly the fuzzy dilation of A by ν, where ν is the fuzzy structuring element defined on $\mathcal{S}$ as:

$$\forall P \in \mathcal{S}, \ \nu(P) = f[\beta(O, P)], \qquad (54)$$

with O as the center of the structuring element. For the linear function used in our experiments, the structuring element is:

$$\forall P \in \mathcal{S}, \ \nu(P) = \max[0, 1 - \frac{2}{\pi} \arccos \frac{\boldsymbol{OP} \cdot \boldsymbol{u}_\alpha}{\|\boldsymbol{OP}\|}]. \qquad (55)$$

It is represented in 2D in Figure 6 (right).

Among the nice properties of this definition is invariance with respect to geometrical transformation (translation, rotation, scaling), which are requirements in object recognition. It also has a behavior that fits well the intuition if the distance to the reference object increases, and in case of concavities. These properties are detailed in [6], and several examples are shown. From an algorithmic point of view, an approximate method, based on propagation algorithms is proposed in [6], that reduces considerably the computation cost and the complexity.

4.6   Example on Brain Structures

In this Section we illustrate the knowledge representation method on a simple example on brain structures. This illustration comes from an atlas-based recognition method described in [27], that uses the type of knowledge representation formalized in this paper. A slice extracted from the atlas 3D volume is presented in Figure 7 (left); the right view shows the corresponding slice in a 3D magnetic resonance image (MRI) to be processed. The labeled image constitutes the iconic part of the model. The propositional part is constituted by expert knowledge about relationships between objects and expected radiometry of each structure. A spatial representation of this knowledge is derived, as presented in the previous sections.

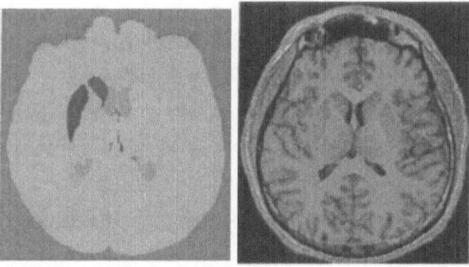

Fig. 7. Slice extracted from a model atlas and from a MRI image. In the atlas, each grey level represents a different object we are interested in.

For instance, the recognition of a caudate nucleus in a 3D MRI image uses previous recognition of brain and lateral ventricles and following knowledge, illustrated in Figure 8:

- rough shape and localization are provided by the representation of the caudate nucleus in the atlas, and its fuzzy dilation to account for variability and for inexact matching between the model and the image,
- the caudate nucleus belongs to the brain (black) but is outside from both lateral ventricles (white components inside the brain),
- the caudate nucleus is lateral to the lateral ventricle.

These pieces of knowledge can be combined (also with information extracted from the image itself), which leads to a successful recognition of the caudate nucleus (see [27] for the fusion and recognition method). Figure 9 illustrates the spatial representation of some knowledge about distances. Figure 10 shows 3D views of some cerebral objects as defined in the atlas and as recognized in an MR image with our method [27]. They are correctly recognized although the size, the location and the morphology of these objects in the image significantly differ from their definitions in the atlas. Note in particular the good recognition of third and fourth ventricles, that are very difficult to segment directly from the image. Here the help of relationships to other structures is very important and conditions the quality of the results.

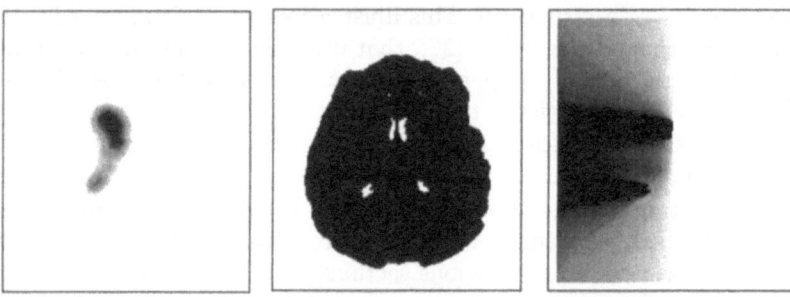

Fig. 8. Information representation in the image space (only one slice of the 3D volume is shown), illustrating knowledge about one caudate nucleus: shape information (left), set relationships (middle), and relative directional relationship (right). Membership values vary from 0 (white) to 1 (black).

5   Symbolic Representations of Spatial Relationships

In this Section, we use the logical framework presented in Section 2.3. For spatial reasoning, interpretations can represent spatial entities, like regions of the space. Formulas then represent combinations of such entities, and define regions, objects, etc., which may be not connected. For instance, if a formula φ is a symbolic representation of a region X of the space, it can be interpreted for instance as "the object we are looking at is in X". In an epistemic interpretation, it could represent the belief of an agent that the object is in X[3]. The interest of such representations could be also to deal with any kind of spatial entities, without referring to points. Using these interpretations, if φ represents some knowledge or belief about a region X of the space, then $\Box\varphi$ represents a restriction of X. If we are looking at an object in X, then $\Box\varphi$

---

[3] This epistemic interpretation is due to Alessandro Saffiotti (personal communication).

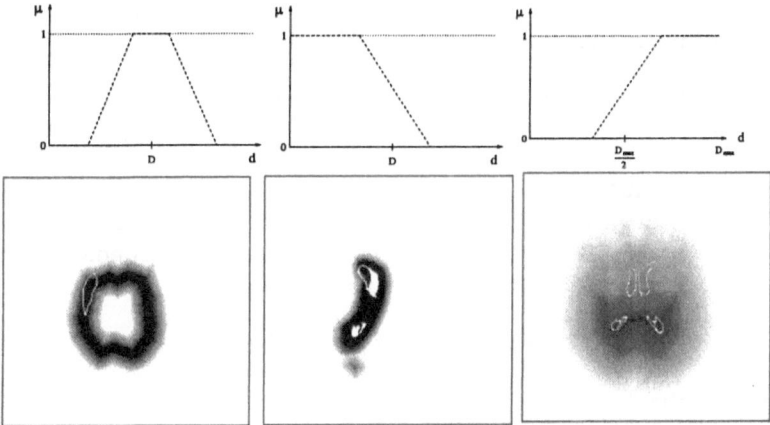

Fig. 9. Examples of representation of knowledge about distances. Top: membership functions μ_n. Bottom: spatial fuzzy sets. The following types of knowledge are illustrated: the putamen has an approximately constant distance to the brain surface (left), the caudate nucleus is at a distance about less than D from the lateral ventricles (in white) (middle), lateral ventricles are inside the brain and at a distance larger than about D from the brain surface (right). The contours of the objects we are looking at are shown in white.

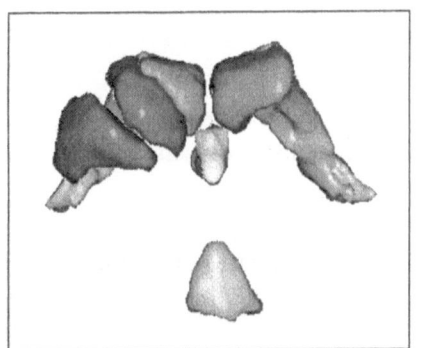 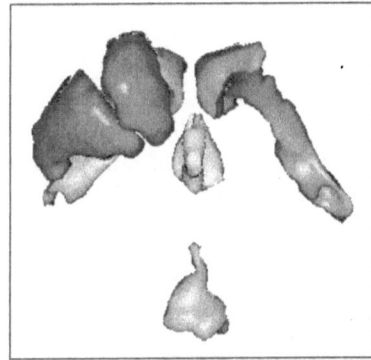

Fig. 10. Recognition results. The left view represents six objects from the model atlas: lateral ventricles (medium grey), third and fourth ventricles (light grey), caudate nucleus and putamen (dark grey). The right view represents the equivalent objects recognized from a MRI acquisition. From [27].

is a necessary region for this object. Similarly, $\Diamond \varphi$ represents an extension of X, and a possible region for the object.

We propose here to use the modal operators introduced in Section 2.3 to provide symbolic and qualitative representations of spatial knowledge.

5.1   Topological Relationships

Let us first consider topological relationships. Let us consider two formulas φ and ψ representing two regions X and Y of the space. Note that all what follows holds in both crisp and fuzzy cases. Simple topological relations such as inclusion, exclusion, intersection do not call for more operators than the standard ones of propositional logic (see e.g. [4]). But other relations such that X is a tangential part of Y can benefit from the morphological modal operators. Such a relationship can be expressed as:

$$\varphi \to \psi \text{ and } \Diamond\varphi \wedge \neg\psi \text{ consistent,} \tag{56}$$

or, equivalently,

$$\varphi \to \psi \text{ and } \varphi \wedge \neg\Box\psi \text{ consistent.} \tag{57}$$

Indeed, if X is a tangential part of Y, it is included in Y but its dilation is not, and equivalently it is not included in the erosion of Y, as illustrated in Figure 11.

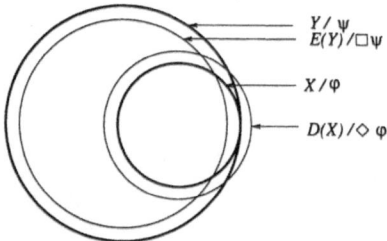

Fig. 11. Illustration of tangential part relationship, and its expression in terms of dilation and erosion.

In a similar way, a relation such that X is a non tangential part of Y is expressed as:

$$\varphi \to \psi \text{ and } \Diamond\varphi \to \psi, \tag{58}$$

or, equivalently,

$$\varphi \to \psi \text{ and } \varphi \to \Box\psi. \tag{59}$$

If we also want X to be a proper part, we have to add the following condition:

$$\neg\varphi \wedge \psi \text{ consistent.} \tag{60}$$

Let us now consider adjacency (or external connection). Saying that X is adjacent to Y means that they do not intersect and as soon as one region is dilated, it has a non empty intersection with the other. In symbolic terms, this relation can be expressed as:

$$\varphi \wedge \phi \text{ inconsistent and } \Diamond\varphi \wedge \psi \text{ consistent and } \varphi \wedge \Diamond\psi \text{ consistent.} \tag{61}$$

Actually, this expression is valid in a discrete domain. If φ and ψ represent spatial entities in a continuous spatial domain, some problems may occur if these entities are closed sets and have parts of local dimension less than the dimension of the space (see [11] for a complete discussion). Such problems can be avoided if the entities are reduced to regular ones, i.e. that are equal to the closure of their interior. Using the topological interpretation of modal logic, this can be expressed as $\varphi \leftrightarrow \Diamond\Box\varphi$.

It could be interesting to link these types of representations with the ones developed in the community of mereology and mereotopology, where such relations are defined respectively from parthood and connection predicates [2,45,16,54]. Interestingly enough, erosion is defined from inclusion (i.e. a parthood relationship) and dilation from intersection (i.e. a connection relationship). Some axioms of these domains could be expressed in terms of dilation. For instance from a parthood postulate $P(X,Y)$ between two spatial entities X and Y and from dilation, tangential proper part could be defined as $TPP(X,Y) = P(X,Y) \wedge \neg P(Y,X) \wedge \neg P(D(X),Y)$.

5.2   Distances

Again we use expressions of minimum and Hausdorff distances in terms of morphological dilations. The translation into a logical formalism is straightforward. Expressing that $d_{\min}(X,Y) = n$ leads to:

$$\begin{cases} \forall m < n, \Diamond^m\varphi \wedge \psi \text{ inconsistent and } \Diamond^m\psi \wedge \varphi \text{ inconsistent} \\ \text{and } \Diamond^n\varphi \wedge \psi \text{ consistent and } \Diamond^n\psi \wedge \varphi \text{ consistent.} \end{cases} \quad (62)$$

Expressions like $d_{\min}(X,Y) \leq n$ translate into:

$$\Diamond^n\varphi \wedge \psi \text{ consistent and } \Diamond^n\psi \wedge \varphi \text{ consistent.} \quad (63)$$

Expressions like $d_{\min}(X,Y) \geq n$ translate into:

$$\forall m < n, \Diamond^m\varphi \wedge \psi \text{ inconsistent and } \Diamond^m\psi \wedge \varphi \text{ inconsistent.} \quad (64)$$

Expressions like $n_1 \leq d_{\min}(X,Y) \leq n_2$ translate into:

$$\begin{cases} \forall m < n_1, \Diamond^m\varphi \wedge \psi \text{ inconsistent and } \Diamond^m\psi \wedge \varphi \text{ inconsistent} \\ \text{and } \Diamond^{n_2}\varphi \wedge \psi \text{ consistent and } \Diamond^{n_2}\psi \wedge \varphi \text{ consistent.} \end{cases} \quad (65)$$

The proof of these equations involves mainly **T** and the last property of Theorem 2 (see Section 2.3).

Similarly for Hausdorff distance, we translate $d_{\text{Haus}}(X,Y) = n$ by:

$$\begin{cases} \forall m < n, \psi \wedge \neg\Diamond^m\varphi \text{ consistent or } \varphi \wedge \neg\Diamond^m\psi \text{ consistent} \\ \text{and } \psi \to \Diamond^n\varphi \text{ and } \varphi \to \Diamond^n\psi. \end{cases} \quad (66)$$

The first condition corresponds to $d_{\text{Haus}}(X,Y) \geq n$ and the second one to $d_{\text{Haus}}(X,Y) \leq n$.

Let us consider an example of possible use of these representations for spatial reasoning. If we are looking at an object represented by ψ in an area which is at a distance in an interval $[n_1, n_2]$ of a region represented by φ, this corresponds to a minimum distance greater than n_1 and to a Hausdorff distance less than n_2. This is illustrated in Figure 12.

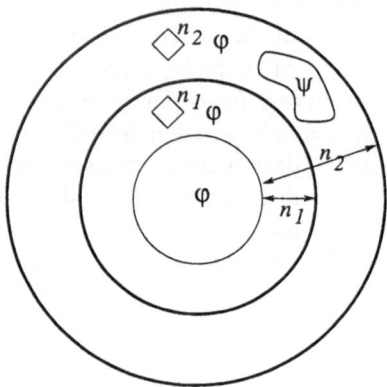

Fig. 12. Illustration of a distance relation expressed by an interval.

Then we have to check the following relations:

$$\psi \to \neg\Diamond^{n_1}\varphi \wedge \Diamond^{n_2}\varphi, \tag{67}$$

or equivalently:

$$\psi \to \Box^{n_1}\neg\varphi \wedge \Diamond^{n_2}\varphi. \tag{68}$$

This expresses in a symbolic way an imprecise knowledge about distances represented as an interval. If we consider a fuzzy interval, this extends directly by means of fuzzy dilation.

These expressions show how we can convert distance information, which is usually defined in an analytical way, into algebraic expressions through mathematical morphology, and then into logical expressions through morphological expressions of modal operators.

5.3   Directional Relative Position

Here we rely again on the approach we proposed in [6], where the reference object is dilated with a particular structuring element, of radial form, having high membership values along lines in the desired direction, and decreasing membership values when going away from this direction.

Let us denote by D^d the dilation corresponding to a directional information in the direction d, and by $\Diamond^d$ the associated modal operator. Expressing that an object represented by ψ has to be in direction d with respect to a region represented by φ amounts to check the following relation:

$$\psi \to \diamond^d \varphi. \tag{69}$$

In the fuzzy case, this relation can hold to some degree.

This formulation directly inherits the properties of directional relative position defined from dilation (see [7] for details), such as invariance with respect to geometrical transformations. It also has a behavior that fits well the intuition if the distance to the reference object increases, and in case of concavities.

6 Conclusion

The spatial arrangement of objects in images provides important information for recognition and interpretation tasks, in particular when the objects are embedded in a complex environment like in medical or remote sensing images. Such information can be expressed in different ways varying from purely quantitative and precise ones to purely qualitative and symbolic ones. We have shown in this paper that mathematical morphology provides a unified and consistent framework to express different types of spatial relationships and to answer different questions about them, with good properties. Due to the strong algebraic structure of this framework, it applies to objects represented as sets, as fuzzy sets, and as logical formulas as well. This establishes links between theories that were so far disconnected. Usually several relationships have to be used together. This aspect can benefit from the developments in information fusion, both in a numerical and in a logical setting, as well as in fuzzy set theory, which was intensively used here to represent spatial imprecision as well as imprecision in expressing knowledge about spatial relations. Applications of this work concern model-based pattern recognition, spatial knowledge representation issues, and spatial reasoning.

References

1. J. Allen. Maintaining Knowledge about Temporal Intervals. *Comunications of the ACM*, 26(11):832–843, 1983.
2. N. Asher and L. Vieu. Toward a Geometry of Common Sense: A Semantics and a Complete Axiomatization of Mereotopology. In *IJCAI'95*, pages 846–852, San Mateo, CA, 1995.
3. E. Bengoetxea, P. Larranaga, I. Bloch, and A. Perchant. Solving Graph Matching with EDAs Using a Permutation-Based Representation. In P. Larranaga and J. A. Lozano, editors, *Estimation of Distribution Algorithms: A New Tool for Evolutionary Computation*, chapter 12, pages 239–261. Kluwer Academic Publisher, Boston, Dordrecht, London, 2001.

4. B. Bennett. Modal Logics for Qualitative Spatial Reasoning. *Bulletin of the IGPL*, 4(1):23–45, 1995.
5. I. Bloch. About Properties of Fuzzy Mathematical Morphologies: Proofs of Main Results. Technical report, Télécom Paris 93D023, December 1993.
6. I. Bloch. Fuzzy Relative Position between Objects in Image Processing: a Morphological Approach. *IEEE Transactions on Pattern Analysis and Machine Intelligence*, 21(7):657–664, 1999.
7. I. Bloch. Fuzzy Relative Position between Objects in Image Processing: New Definition and Properties based on a Morphological Approach. *International Journal of Uncertainty, Fuzziness and Knowledge-Based Systems*, 7(2):99–133, 1999.
8. I. Bloch. On Fuzzy Distances and their Use in Image Processing under Imprecision. *Pattern Recognition*, 32(11):1873–1895, 1999.
9. I. Bloch and J. Lang. Towards Mathematical Morpho-Logics. In *8th International Conference on Information Processing and Management of Uncertainty in Knowledge based Systems IPMU 2000*, volume III, pages 1405–1412, Madrid, Spain, jul 2000.
10. I. Bloch and H. Maître. Fuzzy Mathematical Morphologies: A Comparative Study. *Pattern Recognition*, 28(9):1341–1387, 1995.
11. I. Bloch, H. Maître, and M. Anvari. Fuzzy Adjacency between Image Objects. *International Journal of Uncertainty, Fuzziness and Knowledge-Based Systems*, 5(6):615–653, 1997.
12. I. Bloch, C. Pellot, F. Sureda, and A. Herment. Fuzzy Modelling and Fuzzy Mathematical Morphology applied to 3D Reconstruction of Blood Vessels by Multi-Modality Data Fusion. In D. Dubois R. Yager and H. Prade, editors, *Fuzzy Set Methods in Information Engineering: A Guided Tour of Applications*, chapter 5, pages 93–110. John Wiley & Sons, New-York, 1996.
13. G. Borgefors. Distance Transforms in the Square Grid. In H. Maître, editor, *Progress in Picture Processing, Les Houches, Session LVIII, 1992*, chapter 1.4, pages 46–80. North-Holland, Amsterdam, 1996.
14. B. Chellas. *Modal Logic, an Introduction*. Cambridge University Press, Cambridge, 1980.
15. E. Clementini and O. Di Felice. Approximate Topological Relations. *International Journal of Approximate Reasoning*, 16:173–204, 1997.
16. A. Cohn, B. Bennett, J. Gooday, and N. M. Gotts. Representing and Reasoning with Qualitative Spatial Relations about Regions. In O. Stock, editor, *Spatial and Temporal Reasoning*, pages 97–134. Kluwer, 1997.
17. B. de Baets. Fuzzy Morphology: a Logical Approach. In B. Ayyub and M. Gupta, editors, *Uncertainty in Engineering and Sciences: Fuzzy Logic, Statistics and Neural Network Approach*, pages 53–67. Kluwer Academic, 1997.
18. T.-Q. Deng and H. Heijmans. Grey-Scale Morphology Based on Fuzzy Logic. Technical Report PNA-R0012, CWI, Amsterdam, NL, 2000.
19. D. Dubois and H. Prade. *Fuzzy Sets and Systems: Theory and Applications*. Academic Press, New-York, 1980.
20. D. Dubois and H. Prade. A Review of Fuzzy Set Aggregation Connectives. *Information Sciences*, 36:85–121, 1985.
21. D. Dubois and H. Prade. Weighted Fuzzy Pattern Matching. *Fuzzy Sets and Systems*, 28:313–331, 1988.

22. S. Dutta. Approximate Spatial Reasoning: Integrating Qualitative and Quantitative Constraints. *International Journal of Approximate Reasoning*, 5:307–331, 1991.
23. J. Freeman. The Modelling of Spatial Relations. *Computer Graphics and Image Processing*, 4(2):156–171, 1975.
24. K. P. Gapp. Basic Meanings of Spatial Relations: Computation and Evaluation in 3D Space. In *12th National Conference on Artificial Intelligence, AAAI-94*, pages 1393–1398, Seattle, Washington, 1994.
25. J. Gasos and A. Ralescu. Using Imprecise Environment Information for Guiding Scene Interpretation. *Fuzzy Sets and Systems*, 88:265–288, 1997.
26. J. Gasós and A. Saffiotti. Using Fuzzy Sets to Represent Uncertain Spatial Knowledge in Autonomous Robots. *Journal of Spatial Cognition and Computation*, 1:205–226, 2000.
27. T. Géraud, I. Bloch, and H. Maître. Atlas-guided Recognition of Cerebral Structures in MRI using Fusion of Fuzzy Structural Information. In *CIMAF'99 Symposium on Artificial Intelligence*, pages 99–106, La Havana, Cuba, March 1999.
28. H. J. A. M. Heijmans and C. Ronse. The Algebraic Basis of Mathematical Morphology – Part I: Dilations and Erosions. *Computer Vision, Graphics and Image Processing*, 50:245–295, 1990.
29. G. E. Hughes and M. J. Cresswell. *An Introduction to Modal Logic*. Methuen, London, UK, 1968.
30. J. M. Keller and X. Wang. Comparison of Spatial Relation Definitions in Computer Vision. In *ISUMA-NAFIPS'95*, pages 679–684, College Park, MD, September 1995.
31. L. T. Koczy. On the Description of Relative Position of Fuzzy Patterns. *Pattern Recognition Letters*, 8:21–28, 1988.
32. R. Krishnapuram, J. M. Keller, and Y. Ma. Quantitative Analysis of Properties and Spatial Relations of Fuzzy Image Regions. *IEEE Transactions on Fuzzy Systems*, 1(3):222–233, 1993.
33. B. Kuipers. Modeling Spatial Knowledge. *Cognitive Science*, 2:129–153, 1978.
34. B. J. Kuipers and T. S. Levitt. Navigation and Mapping in Large-Scale Space. *AI Magazine*, 9(2):25–43, 1988.
35. J. Liu. A Method of Spatial Reasoning based on Qualitative Trigonometry. *Artificial Intelligence*, 98:137–168, 1998.
36. J.-F. Mangin, I. Bloch, J. Lopez-Krahe, and V. Frouin. Chamfer Distances in Anisotropic 3D Images. In *EUSIPCO 94*, pages 975–978, Edinburgh, UK, September 1994.
37. G. Matheron. *Eléments pour une théorie des milieux poreux*. Masson, Paris, 1967.
38. G. Matheron. *Random Sets and Integral Geometry*. Wiley, New-York, 1975.
39. P. Matsakis and L. Wendling. A New Way to Represent the Relative Position between Areal Objects. *IEEE Trans. on Pattern Analysis and Machine Intelligence*, 21(7):634–642, 1999.
40. K. Miyajima and A. Ralescu. Spatial Organization in 2D Segmented Images: Representation and Recognition of Primitive Spatial Relations. *Fuzzy Sets and Systems*, 65:225–236, 1994.
41. A. Perchant and I. Bloch. Fuzzy Morphisms between Graphs. *Fuzzy Sets and Systems*, 2001.

42. A. Perchant, C. Boeres, I. Bloch, M. Roux, and C. Ribeiro. Model-based Scene Recognition Using Graph Fuzzy Homomorphism Solved by Genetic Algorithm. In *GbR'99 2nd International Workshop on Graph-Based Representations in Pattern Recognition*, pages 61–70, Castle of Haindorf, Austria, 1999.

43. D. J. Peuquet. Representations of Geographical Space: Toward a Conceptual Synthesis. *Annals of the Association of American Geographers*, 78(3):375–394, 1988.

44. D. Pullar and M. Egenhofer. Toward Formal Definitions of Topological Relations Among Spatial Objects. In *Third Int.. Symposium on Spatial Data Handling*, pages 225–241, Sydney, Australia, August 1988.

45. D. Randell, Z. Cui, and A. Cohn. A Spatial Logic based on Regions and Connection. In *Principles of Knowledge Representation and Reasoning KR'92*, pages 165–176, San Mateo, CA, 1992.

46. A. Rosenfeld. Fuzzy Digital Topology. *Information and Control*, 40:76–87, 1979.

47. A. Rosenfeld. The Fuzzy Geometry of Image Subsets. *Pattern Recognition Letters*, 2:311–317, 1984.

48. A. Rosenfeld and A. C. Kak. *Digital Picture Processing*. Academic Press, New-York, 1976.

49. A. Rosenfeld and R. Klette. Degree of Adjacency or Surroundness. *Pattern Recognition*, 18(2):169–177, 1985.

50. J. Serra. *Image Analysis and Mathematical Morphology*. Academic Press, London, 1982.

51. J. Serra. *Image Analysis and Mathematical Morphology, Part II: Theoretical Advances*. Academic Press (J. Serra Ed.), London, 1988.

52. D. Sinha and E. Dougherty. Fuzzy Mathematical Morphology. *Journal of Visual Communication and Image Representation*, 3(3):286–302, 1992.

53. J. K. Udupa and S. Samarasekera. Fuzzy Connectedness and Object Definition: Theory, Algorithms, and Applications in Image Segmentation. *Graphical Models and Image Processing*, 58(3):246–261, 1996.

54. A. Varzi. Parts, Wholes, and Part-Whole Relations: The Prospects of Mereotopology. *Data and Knowledge Engineering*, 20(3):259–286, 1996.

55. L. Vieu. Spatial Representation and Reasoning in Artificial Intelligence. In O. Stock, editor, *Spatial and Temporal Reasoning*, pages 5–41. Kluwer, 1997.

56. L. A. Zadeh. Fuzzy Sets. *Information and Control*, 8:338–353, 1965.

57. L. A. Zadeh. The Concept of a Linguistic Variable and its Application to Approximate Reasoning. *Information Sciences*, 8:199–249, 1975.

Understanding the Spatial Organization of Image Regions by Means of Force Histograms: A Guided Tour

Pascal Matsakis

Computer Engineering and Computer Science Dept.
University of Missouri-Columbia
Columbia, MO 65211, USA
MatsakisP@missouri.edu

Abstract. Understanding the spatial organization of regions in images is a crucial task, essential to many domains of computer vision. The histogram of forces—a quantitative representation of the relative position between two objects—constitutes a powerful tool dedicated to this task. It encapsulates structural information about the objects as well as information about their spatial relationships. Moreover, it offers solid theoretical guarantees and nice geometric properties. Numerous applications have been studied, and new applications continue to be explored. For instance, force histograms can be compared through similarity measures for fuzzy scene matching. They can be used for describing relative positions in terms of spatial relationships modeled by fuzzy relations. They can also be used for scene description, where relative positions are represented by linguistic expressions. This chapter reviews and classifies work on the histogram of forces. It touches topics as varied as human-robot communication and spatial indexing mechanisms for medical image databases.

Keywords. Force histograms, relative positions, spatial relations, shape matching, scene matching, scene description, fuzzy sets, fuzzy logic, pattern recognition, image analysis, computer vision.

1 Introduction

Understanding the spatial organization of regions in images is a crucial task, essential to countless domains of computer vision. The notion of the histogram of forces, which was first introduced in [16] with the aim of providing new definitions of directional relations (such as "to the right of," "above," "to the west of," "behind"), constitutes a powerful tool for accomplishing this task.

A force histogram is a quantitative representation of the relative position between two 2D objects. It encapsulates structural information about the objects as well as information about their spatial relationships. It is sensitive to the shape of the objects, their orientation and their size. It is also sensitive to the distance between them. In fact, the notion of the histogram of forces allows explicit and variable accounting of metric information. Moreover, it offers solid theoretical guarantees and nice geometric properties, ensures fast and efficient processing of vector data as well as of raster data, and enables the handling of fuzzy objects as well as of crisp objects, intersecting objects as well as of disjoint objects, and unbounded objects as well as of bounded objects.

The applications of the histogram of forces are numerous. So far, they seem to fall into three categories. The applications of the first category make "low-level" use of the histogram, i.e., the relative position between two objects is directly represented by the histogram associated with these objects. The typical application consists in comparing histograms through similarity measures for object pair matching. The question is whether a pair of objects (i.e., the two objects and their spatial relationships) corresponds (up to some geometric transformations, like translation, rotation and scaling) to another pair of objects. These two pairs may come from the same image, or from different images. Object pair matching leads to object matching (when the objects in a given pair are the same), shape matching, and scene matching (when many pairs of object pairs are considered). This is discussed in Section 3. The applications of the second category make "intermediate-level" use of the histogram of forces, i.e., the relative position between two objects is described in terms of a few spatial relationships. These relationships are represented by fuzzy spatial relations, and their evaluation relies on the computation of the histogram associated with the objects. The goal then is to assess specific spatial relationships (e.g., "to the right of"), or to compare the relative position of two objects with the relative position of two other objects (knowing that the objects themselves might be all different). This is discussed in Section 4. Finally, the applications of the third category make "high-level" use of the force histogram. The relative position between two objects is represented by words, i.e., linguistic expressions. These expressions are generated from the histogram associated with the objects, typically through the fuzzy spatial relations mentioned above. This is discussed in Section 5. First of all, in the following section, we present the notion of the histogram of forces and its fundamental geometric properties.

2   The Notion of the Histogram of Forces

In this paper, unless otherwise specified, the term "object" denotes a bounded, 2D crisp object (a rigorous definition of this term for the use of histograms of forces is given in [16,21]). We will return to this matter at the end of Section 2.1.

2.1   Description

The relative position of an object A with regard to another object B is represented by a function F^{AB} from $\mathbf{R}$ into $\mathbf{R}_+$. For any direction θ, the value $F^{AB}(\theta)$ is the total weight of the arguments that can be found in order to support the proposition "A is in direction θ of B." More precisely, it is the scalar resultant of elementary forces. These forces are exerted by the points of A on those of B, and each tends to move B in direction θ (Fig. 1). If F^{AB} is defined on $\mathbf{R}$, i.e., if for any θ the scalar resultant $F^{AB}(\theta)$ is finite, then the pair (A,B) is termed F-*assessable* and F^{AB} is called the *histogram of forces associated with (A,B) via F*, or the *F-histogram associated with (A,B)*. The object A is the *argument*, and the object B the *referent*.

Actually, the letter F denotes a numerical function. Let r be a real. If the elementary forces are in inverse ratio to d^r, where d represents the distance between the points considered, then F is denoted by F_r. The F_0-histogram (histogram of constant forces) and F_2-histogram (histogram of gravitational forces) have very different and very interesting characteristics. The former provides a global view of the situation. It considers the closest parts and the farthest parts of the objects equally, whereas the F_2-histogram focuses on the closest parts.

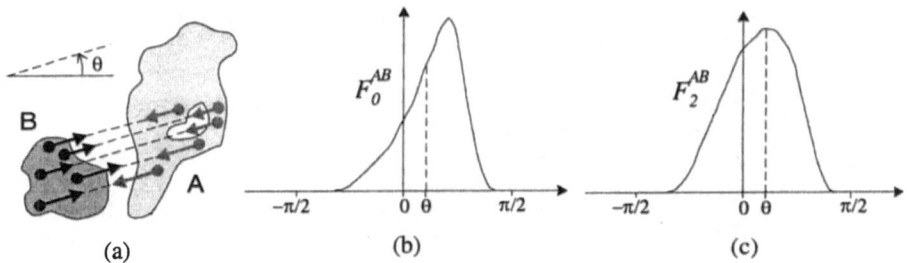

Fig. 1.   Force histograms. (a) $F^{AB}(\theta)$ is the scalar resultant of forces (black arrows). Each one tends to move B in direction θ. (b) The histogram of constant forces associated with (A,B). It represents the position of A relative to B. (c) The histogram of gravitational forces associated with (A,B). It is another representation of the relative position between A and B.

It is shown [16,21] that for any r, any pair of disjoint objects is F_r-assessable. If r is lower than 1, any pair of overlapping objects is F_r-assessable too. The constraint on r can be bypassed by defining histograms of hybrid forces [16,24,33], but then, some geometric properties are lost. As mentioned at the very beginning of Section 2, the term "object" denotes here a bounded, 2D crisp object. Note however that F_r-histograms can also handle unbounded objects (if r is greater than 1 [16,24]), and fuzzy objects (this is discussed in [16,21]). In theory, they can handle neither 0D objects, nor 1D objects. In practice, this is usually not a limitation, since points and lines can easily be assimilated to 2D objects (see, e.g., Fig.

8(e)). Finally, vector data can be processed as well as of raster data [16,21,33]. In the first case (vector data), the complexity of force histogram computation is $O(nN\log(N))$, where N denotes the total number of object vertices and n the number of directions in which forces are computed (usually between 32 and 360, depending on the application that is considered). In the second case (raster data), the complexity is $O(nN\sqrt{N})$, where N denotes the number of pixels of the processed image. This complexity drops to $O(nN)$ for convex objects. Force histogram computation benefits from the power of integral calculus, is highly parallelizable, and utilizes a well-known algorithm that is commonly circuit coded in visualization systems.

2.2 Properties

Force histograms have nice geometric properties. Consider two objects A and B. Assume that (A,B) is F_r-assessable. The following properties hold. Properties 1 to 3 are illustrated by Fig. 2.

Property 1: The pair (B,A) is also F_r-assessable and:
$\forall\theta\in\mathbb{R}$, $F_r^{BA}(\theta) = F_r^{AB}(\theta-\pi)$.

Property 2: Let *sym* be a Δ-axis orthogonal symmetry, and let α be the angle between the X-axis and Δ. The pair $(sym(A), sym(B))$ is F_r-assessable and:
$\forall\theta\in\mathbb{R}$, $F_r^{sym(A)sym(B)}(\theta) = F_r^{AB}(2\alpha-\theta)$.

Property 3: Let *dil* be a central dilation[1] with a positive ratio λ.
The pair $(dil(A), dil(B))$ is F_r-assessable and:
$\forall\theta\in\mathbb{R}$, $F_r^{dil(A)\,dil(B)}(\theta) = \lambda^{3-r} F_r^{AB}(\theta)$.

Property 4: Let *stre* be an X-axis orthogonal stretch with a positive ratio k. For any real θ, let $\overline{\theta}$ be the value $atan(k^{-1}\tan\theta)$ if $\cos\theta$ is positive, the value θ if $\cos\theta$ is zero, and the value $atan(k^{-1}\tan\theta)+\pi$ otherwise. The pair $(stre(A), stre(B))$ is F_r-assessable and: $\forall\theta\in\mathbb{R}$, $F_r^{stre(A)stre(B)}(\theta) = k^{2-r}[1+(k^2-1)\cos^2\theta]^{(r-1)/2} F_r^{AB}(\overline{\theta})$.

Properties 2 and 3 define the behavior of the F_r-histograms towards any similarity transformation. For instance, Property 2 implies Properties 5 and 6 below. The stretch in Property 4 is particular, since its axis is the X-axis, and its ratio is positive. However, the properties 2 and 4 define the behavior of the F_r-histograms towards any orthogonal one-way stretch. All proofs are in [16] (Chapter 2, Appendix A), and [18]. Note that stretches are not similarity transformations.

Property 5: Let *tran* be a translation. $(tran(A), tran(B))$ is F_r-assessable and:
$\forall\theta\in\mathbb{R}$, $F_r^{tran(A)tran(B)}(\theta) = F_r^{AB}(\theta)$.

Property 6: Let *rot* be a ρ-angle rotation. $(rot(A), rot(B))$ is F_r-assessable and:
$\forall\theta\in\mathbb{R}$, $F_r^{rot(A)rot(B)}(\theta) = F_r^{AB}(\theta-\rho)$.

---

[1] A *dilation* is also known as a *homothecy* (or *homothety*).

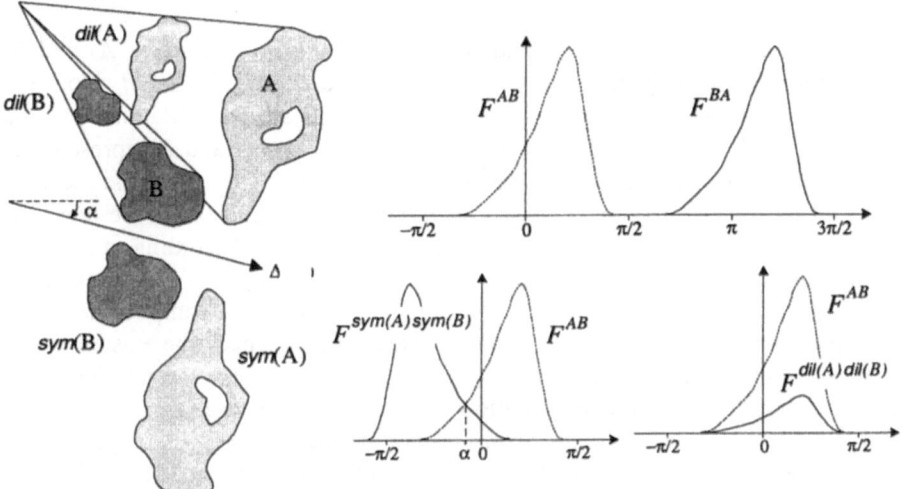

Fig. 2. Some properties of force histograms.
Knowing F^{AB}, it is easy to retrieve F^{BA}, $F^{sym(A)sym(B)}$ and $F^{dil(A)dil(B)}$.

2.3 Inverse Problem

The histogram of forces is sensitive to the shape, the orientation of and the distance between the objects it is associated with. One may then wonder if different pairs of objects can lead to the same histogram. Consider, for instance, two disjoint objects A and B. Let $\mathcal{O}_r^{AB}$ be the set of object pairs (A',B') such that $F_r^{A'B'} = F_r^{AB}$. It is clear that (A,B) belongs to $\mathcal{O}_r^{AB}$. Moreover, it is not the only element of $\mathcal{O}_r^{AB}$. Consider any translation *tran*, any π-angle rotation *rot*, and any dilation *dil*. According to Properties 1, 3, 5 and 6, the pairs $(tran(A), tran(B))$ and $(tran(rot(B)), tran(rot(A)))$ also belong to $\mathcal{O}_r^{AB}$; if r is equal to 3, the pairs $(tran(dil(A)), tran(dil(B)))$ and $(tran(rot(dil(B))), tran(rot(dil(A))))$ belong to $\mathcal{O}_r^{AB}$ too. Does $\mathcal{O}_r^{AB}$ contain other elements than these ones? It is an intricate problem that remains to be solved. However, in practice, if two object pairs (A,B) and (A',B') are such that $F_r^{A'B'}$ is equal to F_r^{AB}, then (A',B') is most probably one of the pairs listed above. A more detailed discussion on this topic can be found in [18].

3   Comparing Force Histograms

Some applications of the histogram of forces make "low-level" use of the histogram. In these applications, histograms are compared through similarity measures.

3.1 Principle

Consider four objects A_1, B_1, A_2 and B_2 in the Euclidean plane (e.g., A_1 and B_1 come from the segmentation of some digital image I_1, and A_2 and B_2 from the segmentation of another image I_2). Assume there exists a central dilation *dil* with a positive ratio λ such that $A_2=dil(A_1)$ and $B_2=dil(B_1)$ (e.g., I_1 and I_2 represent the same physical objects, but have different scaling factors). Let m_1 be the mean of $F^{A_1B_1}$, let m_2 be the mean of $F^{A_2B_2}$, and let μ be some similarity measure (e.g., the classic sigma-count of the intersection over the union [18]). According to Property 3, the dilation ratio λ (i.e., the scaling factor ratio) is $[m_2/m_1]^{1/(3-r)}$. Moreover, the value $\mu(F^{A_2B_2}, (m_2/m_1)F^{A_1B_1})$ tells us about the validity of the assumption concerning the existence of *dil*. Similarly, Property 6 allows us to check the existence of a rotation *rot* such that $A_2=rot(A_1)$ and $B_2=rot(B_1)$, and to retrieve the rotation angle (or *azimuth difference*). Property 4 allows us to handle the case where the projection plane of the camera (or *image plane*) is not parallel to the observed plane. The declination of the camera platform (or *tilt*) can even be retrieved. Naturally, it is possible to consider combinations of the geometric transformations involved in the different properties. The histogram of forces therefore constitutes a powerful tool for object pair matching. This is thoroughly discussed in [18]. Note that when $A_1=B_1$ and $A_2=B_2$ (the histograms $F^{A_1B_1}$ and $F^{A_2B_2}$ are then called *F-signatures*), object pair matching corresponds to object matching. Hence, the notion of the histogram of forces can also be exploited in pattern recognition and classification problems. This has been illustrated in [23,36,22] (Fig. 3). Finally, the force histogram can be of great use in scene matching, which is one obvious application of object pair matching. In Section 3.2, we examine scene matching in LADAR (Laser Radar) imagery, and exploit the fact that force histograms are able to handle fuzzy objects.

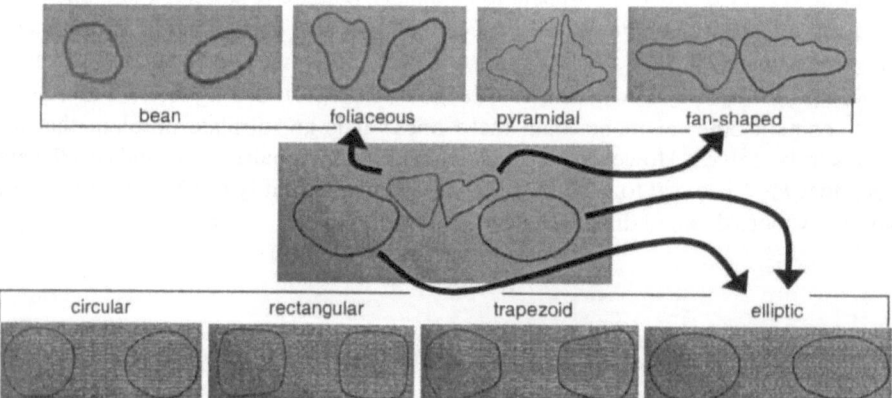

Fig. 3. Experts distinguish four models of sinuses (top) and four models of orbits (bottom). The orbits and sinuses represented by drawings from craniums of the 3rd century A.D. (center) can be classified using the histogram of forces [23,36,22].

3.2   Application to Fuzzy Scene Matching

Consider, for instance, a range image generated by the laser radar system mounted on a surveillance plane. We show in [15] that it is possible to manipulate the three-dimensional data contained within the image and to create a version of the scene as seen from above. In the transformed view, each object is represented by a fuzzy region, and the spatial relationship between two objects is represented by a force histogram. Matching two scenes then comes down to comparing force histograms. Each comparison gives a degree of similarity between the two object pairs under consideration, as well as an assessment of the pose parameters. These values can finally be combined to find the correct scene matching.

First, the range image (Fig. 4(a)) is "lighted." A normal to each pixel is calculated from the three-dimensional positions in the range data of the pixel and its immediate neighbors. It allows an intensity value to be associated with the pixel. The processed scene looks more natural to the human eye (Fig. 4(b)). In [15], a hand segmentation is then performed (Fig. 4(c)). It results in a rough approximation of the objects' edges. The inaccuracies in the segmentation are corrected using range information: each approximated edge is automatically adjusted to correlate with the best possible "real" edge on the range data.

The tilt of the camera platform is assessed using a Hough-like transform. Two consecutive data points in any of the columns of the range data allow a candidate tilt angle to be produced. The most commonly found angle is assumed to be the actual tilt (thus, the method only works on data which represents a relatively flat landscape with objects). A similar method can be used to find the roll of the scene.

The labels of the segmented image are then mapped to the three-dimensional positions of the range data, and these positions are rotated by the tilt angle. The resulting image is the scene as viewed from above (Fig. 4(d)). It has many gaps in it. Some are due to the general spreading of the pixels caused by the rotation. They are easy to handle. The other gaps correspond to uncertainty areas, i.e., areas which were obscured by objects in the original image. They are "filled" by associating each object with a fuzzy region (Fig. 4(d)). To best represent the uncertainty, different types of boundaries are considered, depending on whether the gap occurs on the front side of the object (i.e., the side closest to the camera), or on the back side of it.

Once the two scenes to match have been transformed to a declination independent angle, a force histogram is calculated for each individual fuzzy region pair. At this point, when comparing two histograms not coming from the same scene, the rotational (i.e., azimuth) difference and scaling ratio of the images can be assessed and varied to maximize a given similarity measure. Rotational difference is calculated as the distance between the centroids of the two histograms, and is varied by shifting one of the histograms along the horizontal axis. The scaling ratio is calculated by comparing the histogram averages, and is varied by stretching one of the histograms vertically. This is justified by properties 6 and 3.

In [15], experiments were conducted on a pair of LADAR range images provided by the Naval Air Warfare Center. The two images represent a power-plant complex seen from two different viewpoints. As illustrated by Fig. 5, we focused on a set of four buildings. Our goal was to coherently label the buildings, and to retrieve the declination angles and the rotational difference. Twelve histograms were computed (6 per scene), and 6!=720 possible scene matches were considered. For each possible matching, an overall "matching degree" was derived from the computed maximum similarity measures, scaling ratios, and rotational differences. The

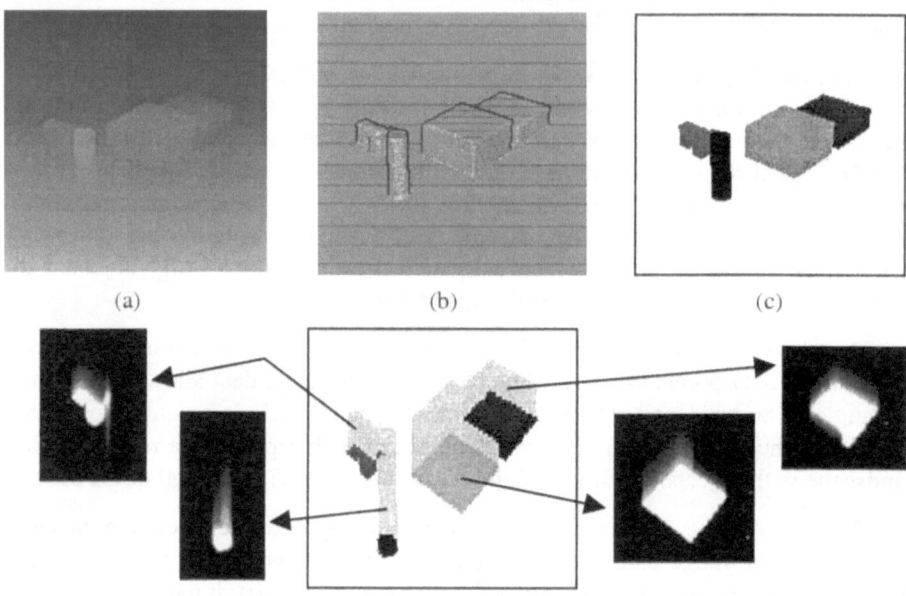

Fig. 4. By manipulating the three-dimensional data contained within a range image, it is possible to create a version of the scene as seen from above [15]. In the transformed view, each object is represented by a fuzzy region. (a) Range image. (b) Lighted scene. (c) Segmented image. (d) Overhead view and fuzzy regions.

Fig. 5. These two images represent a power-plant complex seen from two different viewpoints. A fuzzy scene matching approach based on force histogram computation makes it possible to coherently label the buildings, and to retrieve the declination angles and the rotational difference.

matching with the highest degree was found to be the true matching and led to a coherent labeling of the buildings. The actual pose parameters were not provided by the NAWC and could not be compared with the recovered values. However, extensive experiments on synthetic data have shown that the tilts can be recovered to within 5° and the azimuth difference to within 10° [15,18].

4   Defining Fuzzy Spatial Relations

In [5], Freeman proposed that the relative position of two objects be described in terms of spatial relationships. He also proposed that fuzzy relations be used, because "all-or-nothing" standard mathematical relations are clearly not suited to models of spatial relationships. Freeman's ideas were widely adopted. But many authors assimilated 2D objects to very elementary entities such as a point (centroid) or a (bounding) rectangle. This approach is extremely practical, therefore it has often been used, notably for spatial reasoning and representation and processing of qualitative spatial knowledge. However, it cannot be hoped to give a satisfactory modeling of the relationships, because a lot of morphological information on the considered objects is lost. We show here that the histogram of forces— which encapsulates a large amount of information on the objects[2]—lends itself, with great flexibility, to the definition of fuzzy spatial relations. This is the "intermediate-level" use of the force histogram.

4.1   Directional Relations

Relative position is often assimilated to directional relationships. The point of view is improper, but it shows the importance of these relationships in the field of computer vision. A *family of fuzzy directional spatial relations* is a series $(\mathcal{R}_\alpha)_{\alpha \in \mathbb{R}}$ of fuzzy binary relations. The relation $\mathcal{R}_\alpha$ reads "in direction α of." Depending on α and on context, it may also read "to the right of," "above," "to the west of," "in front-left of," etc. It connects any pair of spatial entities A and B with a numerical value[3]. This value, $A\mathcal{R}_\alpha B$, corresponds to the degree of truth of the proposition "A is in direction α of B." It is a real number greater than or equal to 0 (proposition completely false) and less than or equal to 1 (proposition completely true). Entity A is the *argument* of the proposition and entity B the *referent*. Points are easy to handle (Fig. 6), but the problem gets complex when parameters such as shape, size and orientation are involved [29,25]. Although numerous methods for defining families of directional relations between 2D objects can be found in the literature, few of these methods simultaneously meet the following requirements:

---

[2] The histogram of forces does not constitute, of course, the only way to preserve structural information (see, for instance, [2] and [25]).

[3] The pair may also be connected with a confidence interval or a fuzzy number.

(a) No object is assimilated to an elementary entity such as a point or a rectangle.

(b) The defined relations are fuzzy relations, and not "all-or-nothing" ones.

(c) The defined family satisfies the basic axiomatic properties [16,21] which are—in a more or less explicit way—widely adopted by computer scientists: *(i)* two objects can be assimilated to points if they are distant enough; *(ii)* the directional relations are not sensitive to scale; *(iii)* neither a space dimension nor a direction are preferred; *(iv)* the semantic inverse principle [5] is respected (e.g., *A* is to the left of *B* as much as *B* is to the right of *A*).

The centroid method (see, e.g., [10]) and the methods described in [7,9] do not meet requirement **(a)**; the ones described in [1,6,13] do not meet requirement **(c)**. Actually, as far as we are aware, the only methods that fairly meet the previous requirements are based—explicitly or not—on the notion of the histogram of angles presented in [25]. These methods are the compatibility method [25], the aggregation method [14], the possibility method proposed in [2][4] (but not the necessity method, neither the average one), and, to a certain extent, the neural network methods [11]. In [16,19], we showed that the corresponding families of directional relations can be advantageously redefined using force histograms instead of angle histograms. In [16,20], we noted that most families of fuzzy relations run counter to the fact that, generally, people do not combine more than two spatial prepositions when translating visual information into natural language descriptions [8,27]. We also exhibited a coherent and rational perception of the world that no existing family could model.

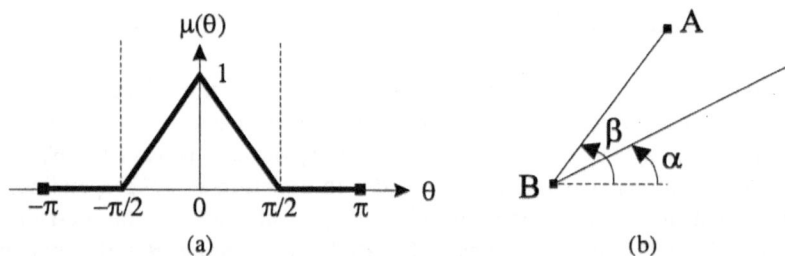

Fig. 6. Example of directional relations between points. (a) A typical fuzzy set. (b) The degree of truth of the proposition "*A* is in direction α of *B*" is $\mu(\beta-\alpha)$.

These facts led us to introduce alternative families based on the notion of the histogram of forces [16,20]. The idea is to impose physical considerations on the histograms. Let *r* be a real and *(A,B)* an F_r-assessable pair of objects. Our goal is to assess the degree of truth of a proposition like "*A* is in direction α of *B*," where α represents any angle. Here, we will only consider the proposition "*A* is in direction 0 of *B*," which will be read "*A* is to the right of *B*." For another value of α,

---

[4] Although its definition is based on a morphological and fuzzy pattern matching approach, the possibility degree introduced in [2] is basically a function of the histogram of angles.

you can simply perform the computations described below on the shifted histogram, $F_r^{AB}(\theta+\alpha)$. The forces exerted on B are classified in different types. First, the set of directions is divided into four quadrants as shown in Fig. 7. The forces $F_r^{AB}(\theta)$ of the outer quadrants ($\theta \in [-\pi, -\pi/2] \cup [\pi/2, \pi]$) are elements which, to various degrees, weaken the proposition "A is to the right of B"; the forces of the inner quadrants ($\theta \in [-\pi/2, 0] \cup [0, \pi/2]$) are elements which support the proposition. Some forces of the third quadrant are used to compensate—as much as possible—the contradictory forces of the fourth one. The proportion of these compensatory forces is defined by some angle θ_+. Forces of the second quadrant are used in a similar way to compensate the contradictory forces of the first one. The amount of these compensatory forces is defined by θ_-. The remaining forces are called the effective forces. A threshold τ divide them into optimal and sub-optimal components. The optimal components support the idea that A is "perfectly" to the right of B: whatever their direction, they are regarded as horizontal and pointing to the right. The "average" direction $\alpha_r(RIGHT)$ of the effective forces is then computed, in conformity with this agreement.

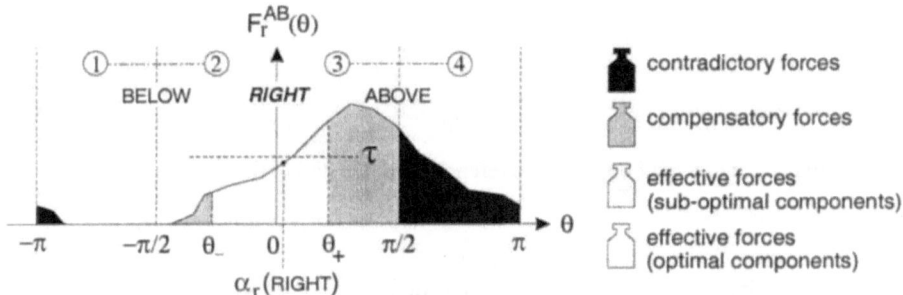

Fig. 7. Force typology associated with the proposition "A is to the right of B."

Finally, the degree of truth of "A is to the right of B" is set to $\mu(\alpha_r(RIGHT)) \times \bar{a}_r(RIGHT)$. In this expression, $\bar{a}_r(RIGHT)$ denotes the percentage of the effective forces (i.e., the sum of the effective forces divided by the sum of all forces), and μ is the membership function of a fuzzy set that can be employed to define a family of fuzzy directional relations between points. In our experiments, we used the typical triangular function graphed in Fig. 6(a). The most optimistic point of view consists in saying that any effective force is optimal, i.e., $\tau=+\infty$. Then, $\mu(\alpha_r(RIGHT)) \times \bar{a}_r(RIGHT)$ is equal to $\bar{a}_r(RIGHT)$—since $\alpha_r(RIGHT)$ is 0 and $\mu(0)$ is 1. The most pessimistic point of view consists in saying that any effective force is sub-optimal, i.e., $\tau=0$. In that case, the expression $\mu(\alpha_r(RIGHT)) \times \bar{a}_r(RIGHT)$ gives some value $\underline{a}_r(RIGHT)$. Setting τ to the average—or a weighted average—of the effective forces constitutes a natural compromise. The degree of truth of "A is to the right of B" is then found to be some value $a_r(RIGHT)$. The 3-tuple ($\underline{a}_r(RIGHT)$, $a_r(RIGHT)$, $\bar{a}_r(RIGHT)$) defines a triangular fuzzy number. It corresponds to the histogram's "opinion" regarding the proposition "A is to the right of B." According to F_r^{AB}, the degree of truth of that proposition is $a_r(RIGHT)$, the maximum degree of truth that can reasonably be attached to it (say, by another source of information) is $\overline{a}_r (RIGHT)$, and the minimum degree that can reasonably be attached to it is $\underline{a}_r(RIGHT)$. The method has been presented in detail in [20]. The French-speaking reader is also invited to consult [19] (or [16]). Fig. 8 shows six pairs of objects. For each pair (A,B), four propositions have been assessed, using two families of fuzzy directional spatial relations. The four propositions are "A is to the right of B," "A is above B," "A is to the left of B" and "A is below B." The two families, **F0** and **F2**, are based on the construction of F_0 and F_2-histograms, and the distinction between contradictory, compensatory and effective forces, as described above. The degrees of truth produced by **F0** and **F2** are displayed in Table 1. Different comparative studies can be found in [16,21,19].

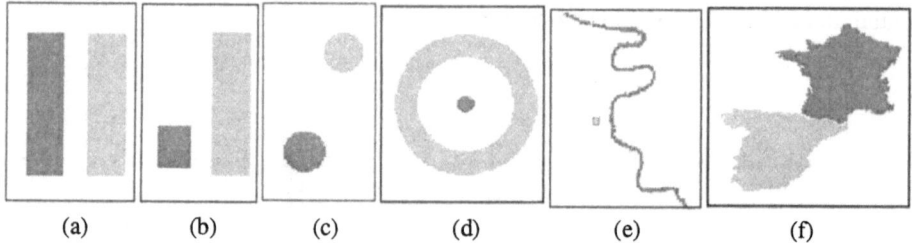

| (a) | (b) | (c) | (d) | (e) | (f) |

Fig. 8. Six pairs of objects. The referent is drawn darker than the argument.

TABLE 1

F0 and **F2**'s opinions regarding the relative position of the objects displayed in Fig. 8 (the degrees of truth are in hundredths). According to the two families, an object cannot be simultaneously a bit to the left and a bit to the right of another. In particular, as illustrated by case (d), directional relationships should not substitute for surroundedness (see Section 4.2.1). Also note that **F2** sometimes conflicts with **F0** (i.e., $a_2 > \overline{a}_0$ or $a_2 < \underline{a}_0$), and vice versa (e.g., case of the house and the river). This can be exploited at a higher level (see Section 5).

F0	$\underline{a}_0$	a_0	$\overline{a}_0$	$\underline{a}_0$	a_0	$\overline{a}_0$	$\underline{a}_0$	a_0	$\overline{a}_0$	$\underline{a}_0$	a_0	$\overline{a}_0$	$\underline{a}_0$	a_0	$\overline{a}_0$	$\underline{a}_0$	a_0	$\overline{a}_0$
RIGHT	100	100	100	69	88	100	23	23	100	0	0	0	0	0	0	0	0	0
ABOVE	0	0	0	31	38	58	77	87	100	0	0	0	0	0	0	0	0	0
LEFT	0	0	0	0	0	0	0	0	0	0	0	0	82	87	92	44	54	83
BELOW	0	0	0	0	0	0	0	0	0	0	0	0	18	19	21	56	76	99
	(a)			(b)			(c)			(d)			(e)			(f)		

F2	$\underline{a}_2$	a_2	$\overline{a}_2$	$\underline{a}_2$	a_2	$\overline{a}_2$	$\underline{a}_2$	a_2	$\overline{a}_2$	$\underline{a}_2$	a_2	$\overline{a}_2$	$\underline{a}_2$	a_2	$\overline{a}_2$	$\underline{a}_2$	a_2	$\overline{a}_2$
RIGHT	100	100	100	82	100	100	23	23	100	0	0	0	0	0	0	0	0	0
ABOVE	0	0	0	18	21	37	77	87	100	0	0	0	0	0	0	0	0	0
LEFT	0	0	0	0	0	0	0	0	0	0	0	0	95	98	99	24	29	39
BELOW	0	0	0	0	0	0	0	0	0	0	0	0	5	5	7	76	93	98

4.2 Other Spatial Relations

The histogram of forces was introduced with the aim of providing new definitions of directional relations [16,21], and its use for the modeling of other relationships has been the subject of little investigation. It is clear that the force histogram does not constitute an ideal representation of the relative position between objects, i.e., it does not enable the definition of "any" fuzzy spatial relation. Nevertheless, it can be employed to model a variety of relationships—although it might mean working under certain assumptions on the objects, or using extra geometric features. This is what we illustrate here with two examples.

4.2.1 Surroundedness

Surroundedness can be considered a particular case of *separation* [28]. It is an important spatial relationship in the interpretation of a scene, and many quantitative definitions have been proposed. There are two main approaches. The first approach relies on the fact that according to most families of fuzzy directional relations an object can be in many directions with respect to another. As mentioned in Section 4.1, this feature is questionable: usually, people do not combine more than two spatial prepositions when translating visual information into natural language descriptions [8,27]. However, some authors [25,2] support the idea that it allows "surrounds" (and "is surrounded by") to be derived. Knowing that A is somewhat above, below, to the right and to the left of B as well, one could conclude that A surrounds B. In fact, drawing such a conclusion is not reasonable, unless it is known that the argument A does not intersect the convex hull of B. In other words, "A surrounds B" can be assessed only if it is known that B *does not surround A at all*. The reason is that the directional relations are tied by the *semantic inverse* principle [5] (e.g., A is to the left of B as much as B is to the right of A). Therefore, without constraints on the objects, there is no way to know which one surrounds (or includes!) the other. The second approach derives from Rosenfeld's visual surroundedness [30]. It is based on the computation of a histogram of angles. It supposes that the argument A is connected and does not intersect B. For any pixel P of B, let θ_P be the angle made by the two tangents from P to A as in Fig. 9(a). To each element θ of $[0,2\pi]$, the histogram associates the number of pixels P such that θ_P is equal to θ. In [35], the degree of truth for "A surrounds B" is produced by a multilayer perceptron fed by the histogram values and trained on aggregate responses from a panel of people. Other authors resort to a decreasing membership function μ from $[0,2\pi]$ into $[0,1]$. The function μ is chosen such that $\mu(\theta)$ is 1 if θ is 0, and is 0 if θ is greater than π. In [26], the histogram of angles is assimilated to a fuzzy set and matched to μ, using the compatibility notion [4]. The degree of truth for "A surrounds B" is obtained as the center of gravity of the compatibility fuzzy set. In [14], the histogram is used to compute the aggregated value (e.g., the arithmetic mean, or the generalized mean [12]) of the $\mu(\theta_P)$, when P describes B. The degree of truth for "A surrounds B" is set to this value. Compared with the first one, the second approach gives definitions of surroundedness that are much more consistent with human perception [35]. However, the methods proposed are computationally expensive, and vector data cannot be handled.

We show in [33] that the histogram of forces can easily be employed to assess surroundedness. Let μ be a membership function as above, and let $[\theta_1,\theta_2]$ be the largest interval—with θ_1 in $]-\pi,\pi]$—on which the force histogram F^{AB} is zero. Fig. 9(b) illustrates the meaning of angles θ_1 and θ_2. The degree of truth for "A surrounds B" is set to $\mu(\theta_2-\theta_1)$. Once again, the argument is supposed to be connected. Moreover, it should not intersect the convex hull of the referent (like in the first approach). The definition is quite simple, but compares with the others [33]. Some examples are shown in Fig. 10.

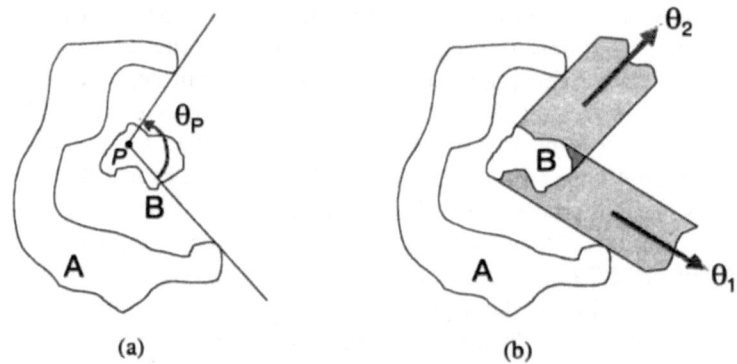

Fig. 9. Defining surroundedness.
(a) Use of a histogram of angles. (b) Use of a histogram of forces.

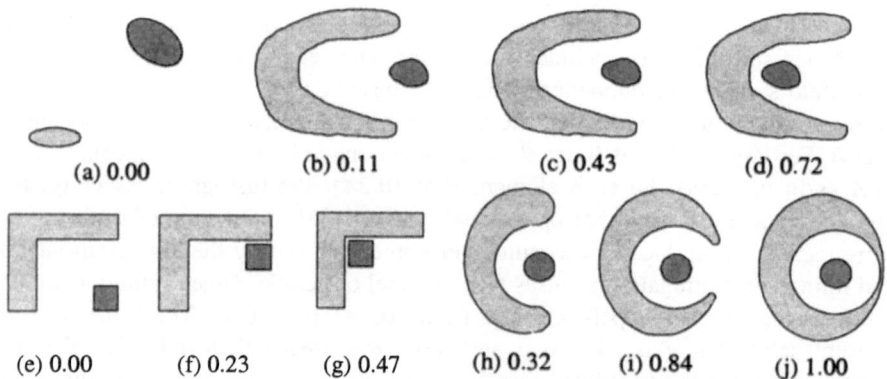

Fig. 10. Surroundedness based on the histogram of forces. In our experiments, μ was linear: for any θ in $[0,\pi]$, $\mu(\theta)=1-\theta/\pi$. The value 1.00 means that the proposition "A surrounds B" is assessed to be completely true, and the value 0.00 that it is completely false (where A denotes the light gray object and B the dark one). Similar images are used in [35] for training and testing the network-based method, and in [33] for comparing different methods.

The advantages of the force histogram-based method are that it ensures faster processing of raster data, and it is able to handle vector data as well. Also, the directional relations can be assessed concurrently. Note that the degree of truth for "A surrounds B" does not really depend on the histogram values. The only thing that matters is which values are equal to zero and which ones are not. This has two important consequences. First, the results do not depend on the choice of the force histogram. Second, the method is not extremely robust. Slight changes in the object shapes (especially at the two "ends" of the argument) may have a noticeable impact on the degrees of truth. In fact, similar comments can be addressed to any method that derives from Rosenfeld's visual surroundedness: the results are not sensitive to the thickness of the argument, only to tangency points. The force histogram-based definition is exploited in Section 5.2.2. The application described there does not suffer from the above-mentioned limitations (constraints on the objects, low robustness). However, other applications might. In [17], the degree of truth for "A surrounds B" is redefined using the force histogram values (and not only the fact that these values are either zero or non-zero). Related relationships, like "between" and "among," are also examined. Designing a new type of histogram of forces constitutes another promising avenue. The idea would be to adopt a novel set of axiomatic properties, and to change the way the longitudinal sections are handled [16,21].

4.2.2   Inner-Adjacency

Adjacency is another important spatial relationship between image regions. Different quantitative definitions have been proposed, notably in [37,30,3]. Our work on spatial indexing mechanisms for medical image databases has led us to consider a particular relationship called "inner-adjacency." In [31], the position of an object A relative to another object B is represented by the histogram of constant forces associated with $(A,B-A)$. Some histogram values may thus be zero even when A and B intersect. The degree of truth of the proposition "A is inner-adjacent to B" is set to

$$max\left(\frac{i}{b}, min\left(\frac{i}{a}, 1-\frac{hmin}{hmean}\right)\right),$$

where a denotes the size (area) of the argument, b the size of the referent, i the size of $A \cap B$, and $hmin$ and $hmean$ are the histogram minimum and average values. Fig. 11 shows 8 schematic configurations and the corresponding degrees of truth.

The fuzzy spatial relation presented here models one clinically meaningful relationship. It was implemented in the spatial indexing mechanism of the medical content-based image retrieval system proposed in [31]. The system was tested using a set of 2,080 HRCT lung images (Fig. 12). It achieved a 90% accuracy rate for lesion retrieval based on inner-adjacency. Spatial relationships between lesions and anatomical landmarks in medical images are critically important in disease diagnosis.

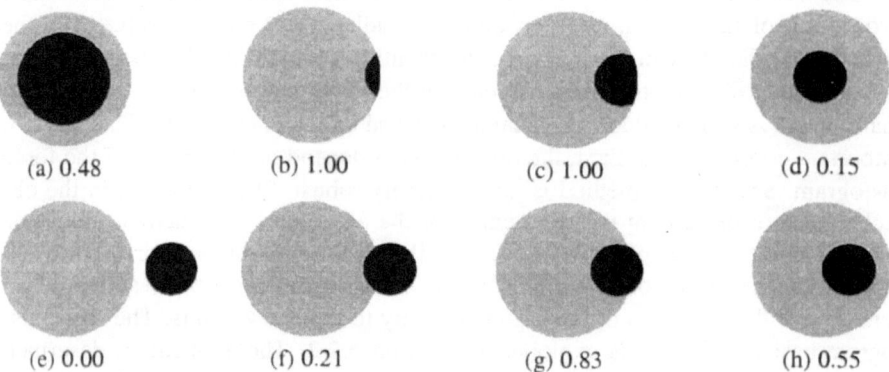

Fig. 11. Inner-adjacency. Each value is the degree of truth of the proposition "*A* is inner-adjacent to *B*," where *A* denotes the black object and *B* the intersected gray one.

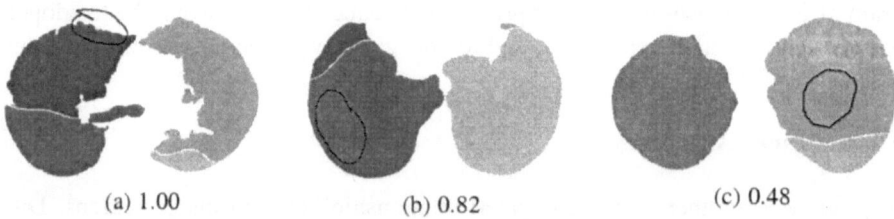

Fig. 12. Lesion retrieval based on inner-adjacency. The three images were generated from HRCT (High Resolution Computed Tomography) lung images. Consider (a). The two lungs are clearly visible. Fissures divide them into chambers. A lesion has been identified by the physician (note that the region enclosed by the line outside the lung is not part of the lesion). In (a)(b)(c), we are interested in the position of the lesion relative to the chamber it belongs to. The inner-adjacency decreases from left to right.

5   Generating Linguistic Spatial Descriptions

High-level computer vision applications hold a great potential for fuzzy set theory because of its links to natural language. Linguistic scene description, a language-based interpretation of regions and their relationships, is one such application that is starting to bear the fruits of fuzzy set theoretic involvement. In this section, we show how the fuzzy relations presented in Section 4 can be utilized to produce logical linguistic spatial descriptions along with assessments as to the validity of the descriptions. This is the "high-level" use of the histogram of forces.

5.1   Principle

In [20], a linguistic description of the relative position between any 2D objects A and B is generated from F_0^{AB} (the histogram of constant forces associated with (A,B)) and F_2^{AB} (gravitational forces). As already mentioned in Section 2.1, these two histograms have very different and very interesting characteristics. The former, F_0^{AB}, provides a global view of the situation. It considers the closest parts and the farthest parts of the objects equally, whereas F_2^{AB} focuses on the closest parts.

The linguistic description output by the system in [20] relies on the sole primitive directional relationships: "to the right of," "above," "to the left of" and "below" (imagine that the objects are drawn on a vertical surface). First, the histograms' opinions regarding these relationships are computed (see Section 4.1). For instance, the following triangular fuzzy numbers are extracted from F_2^{AB}: $(\underline{a_2}(RIGHT),$ $a_2(RIGHT),$ $\overline{a}_2 (RIGHT)),$ $(\underline{a_2}(ABOVE),$ $a_2(ABOVE),$ $\overline{a}_2 (ABOVE)),$ $(\underline{a_2}(LEFT),$ $a_2(LEFT),$ $\overline{a}_2 (LEFT))$ and $(\underline{a_2}(BELOW),$ $a_2(BELOW),$ $\overline{a}_2 (BELOW))$. For each one of the four primitive directions (say, $RIGHT$), F_0^{AB} may consider that F_2^{AB}'s opinion is defensible $(\underline{a_0}(RIGHT) \leq a_2(RIGHT) \leq \overline{a}_0 (RIGHT))$, or is not defensible $(a_2(RIGHT) < \underline{a_0}(RIGHT)$ or $a_2(RIGHT) > \overline{a}_0 (RIGHT))$, and vice versa. The histograms' opinions are combined, as illustrated in Fig. 13. Six features result from this combination: two primitive directions (a *primary* direction and a *secondary* direction), and four numeric values (a degree of truth and a measure of agreement are associated with each direction). They feed a fuzzy rule base that produces the expected linguistic description. The system handles a set of 16 adverbs (like "mostly," "perfectly," etc.) which are stored in a dictionary, with other terms, and can be tailored to individual users. A description is generally composed of three parts. The first part involves the primary direction (e.g., "A is mostly to the right of B"). The second part supplements the description and involves the secondary direction (e.g., "but somewhat above"). The third part indicates to what extent the four primitive di-

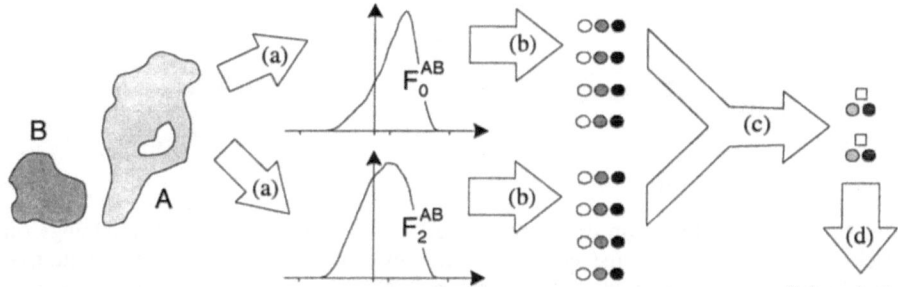

Fig. 13. Generation of linguistic descriptions.
Synoptic diagram of the system presented in [20].
(a) The histograms of constant and gravitational forces are computed. (b) Each histogram gives its opinion about the relative position between the objects. (c) The two opinions are combined. (d) A fuzzy rule base outputs the description.

"A is perfectly to the right of B, but strongly shifted upward. The description is satisfactory."

rectional relationships are suited to describing the relative position of the objects (e.g., "the description is satisfactory"). In other words, it indicates to what extent it is necessary to turn or not to other spatial relations. For instance, if the self-assessment is "not satisfactory," then the system proposed in [33] turns to "surrounds," using the force histogram-based method presented in Section 4.2.1. All details can be found in [20] and [33].

5.2   Application to Image Scene Description

The system for linguistic scene description developed in [20] was tested on numerous synthetic and real data examples. In particular, we used it to describe the relative position between regions from LADAR (Laser Radar) range images of the power-plant at China Lake, CA. These images were provided by the Naval Air Warfare Center. They were processed by applying first a median filter, and then a pseudo-intensity filter. Finally, the filtered images were segmented and labeled manually. Here, contrary to what is said in Section 3.2, range information was not utilized to correct the inaccuracies in the segmentation. Fig. 14 shows some pairs of objects (or groups of objects) that were examined in our experiments.

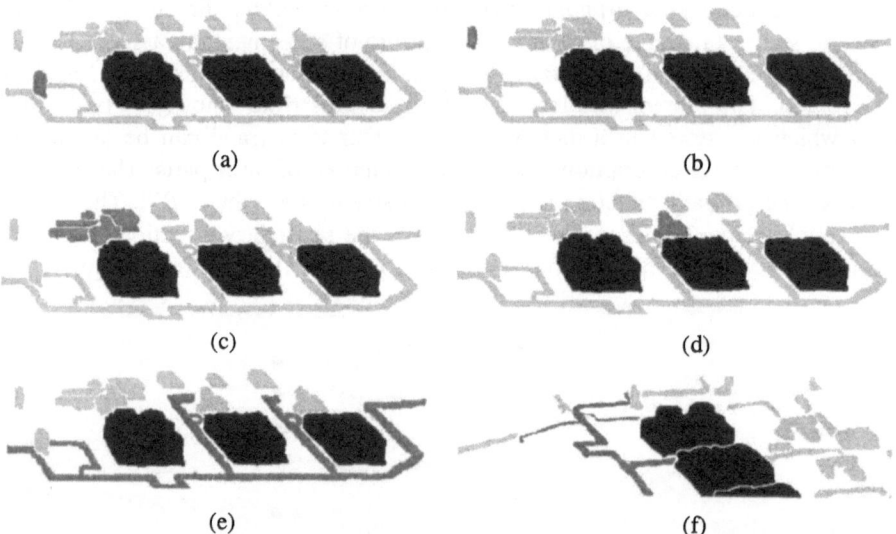

Fig. 14.   (a) "The tower (in dark gray) is perfectly to the left of the stackbuildings (in black). The description is satisfactory." (b) "The tower is to the left of the stackbuildings, but a little above. The description is satisfactory." (c) "The group of storehouses is loosely above-left of the stackbuildings. The description is satisfactory." (d) "The storehouse is perfectly above the stackbuildings, but slightly shifted to the left. The description is satisfactory." (e) "???????" (f) "The pipe is loosely to the left of the stackbuildings, but slightly shifted downward. The description is rather satisfactory."

The richness of the system's language is generally very well employed. Consider Fig. 14(c): the system notes that the relationship is not a perfect above-left, and uses the adverb "loosely" to indicate a bias in one direction. Now, consider Fig. 14(d): the system points out that the storehouse is slightly shifted to the left. In some cases, no pertinent description relying on the sole primitive directional relationships can be given, and the message "???????" is generated. For instance, the relative position between the pipe and the stackbuildings of Fig. 14(e) cannot be described. The output is appropriate, since surroundedness is not considered in [20]. Although the system globally performs very well and produces good intuitive results, some descriptions are not totally satisfactory. Dealing with a rich language is tricky, i.e., it is always easier to be right when vague and imprecise. Consider Fig. 14(f). A piece of the pipe extends between the uppermost and middle stackbuildings. At the end of the extension, the pipe has a strong "downward" relationship with the uppermost building (and a weak "upward" relationship with the middle one). As a result, the argument is assessed to be slightly shifted downward relative to the referent. Note, however, that a good amount of ambiguity is detected, and the system itself considers the description *rather* satisfactory.

5.3   Application to Human-Robot Communication

In [33], we show how linguistic expressions can be generated to describe the spatial relations between a mobile robot and its environment, using readings from a ring of sonar sensors (see also [32] and [34]). Our work is motivated by the study of human-robot communication for non-expert users. The eventual goal is to utilize these linguistic expressions for navigation of the mobile robot in an unknown environment, where the expressions represent the qualitative state of the robot with respect to its environment, in terms that are easily understood by humans. The differences between the systems described in [20] and [33] are few, but they are not inconsiderable. In [20], the force histograms are computed from raster data, and the spatial reference frame is implicitly determined by the reader's location (*world* view). In [33], the histograms are computed from vector data, and the reference frame is determined by the intrinsic orientation of the robot (*egocentric* view). The two works therefore complement one another. They illustrate the fact that the histogram of forces is able to handle vector data as well as of raster data, and makes it easy to switch between a world view and an egocentric view. The system in [33] also considers surroundedness, using the fuzzy relation presented in Section 4.2.1. Since the reference object is always the robot (a Nomad 200 with 16 sonar sensors evenly distributed along its circumference), the limitations mentioned in that section are not an issue (the robot is convex and is not supposed to jam itself in the walls). Moreover, the system expresses proximity information (based on the sonar readings). Two levels of abstraction are provided, and detailed individual descriptions can be combined into more synthetic descriptions. On the other hand, the language used to describe directional positions is coarser than the one in [20] (and the dictionaries are slightly different). The reason is that only a rough representation of the environment objects can be built anyway. Consider Fig. 15(a). The robot is heading rightwards in an angled hallway. There are ten sonar returns (i.e., ten sensors return non-maximum range values). Each one gives a trapezoid, located in the corresponding sonar cone, and built using a constant arbitrary depth. There is a question on whether adjacent sonar readings are from a single object or multiple objects. If the robot cannot fit between the object parts that are responsible for two adjacent sonar readings, then we consider these parts to be from the same obstacle (and the trapezoids are linked). Even if there are actually two objects (this happens in Fig. 15(d), right behind the robot), they may be considered as one for robot navigation purposes. If the robot can fit, we consider separate obstacles (the trapezoids are not linked). The two readings may come from the same object, but there is no way to know that until the robot gets closer and we have a better resolution of the object (since more sensors would

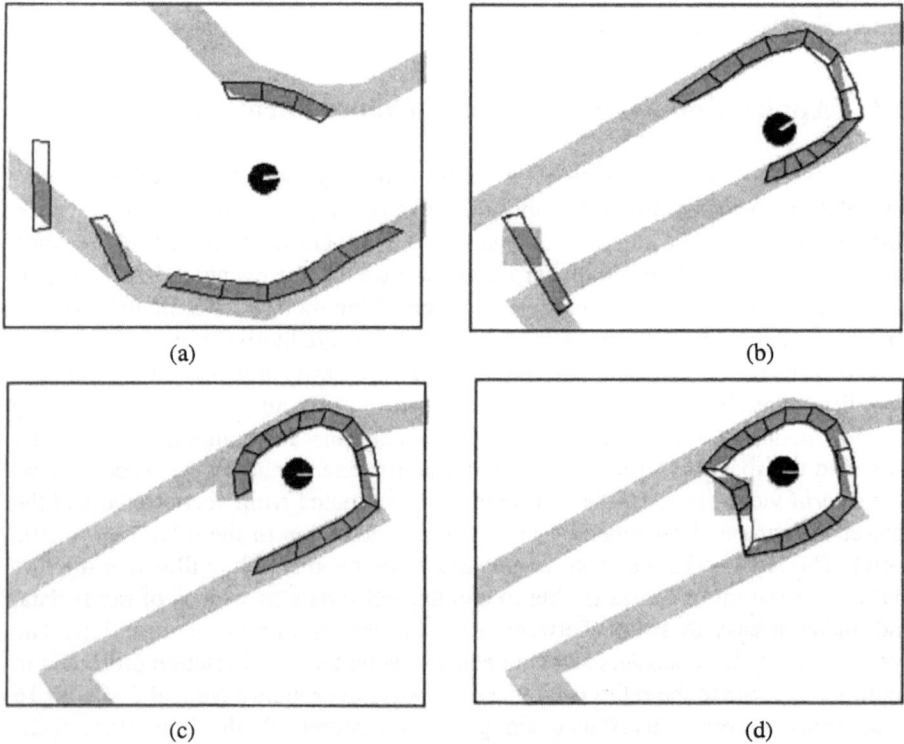

(a)

(b)

(c)

(d)

Fig. 15. A robot describes its environment. (a) "There is an obstacle on my right; it extends forward; it is very close. There is one on my left; it extends forward; it is very close. One is behind me; it is close. Another one is mostly behind me, but somewhat to the right; it is close." (b) "There is an obstacle that surrounds me on the front. There is another one behind me; it is far." (c) "There is an obstacle that surrounds me, but there is an opening on the rear-right." (d) "I am surrounded."

detect its presence). In Fig. 15(a) for instance, two obstacles are detected behind the robot, although the readings come from the same wall. Note that the caption only shows the detailed individual descriptions. In the higher level of abstraction, the robot describes its environment as follows: "There is an obstacle on my right, one on my left, two behind me." The three other figures, Fig. 15(b) to Fig. 15(d), illustrate how the system handles surroundedness, using the fuzzy relation presented in Section 4.2.1.

The experiments were carried out with the Nomad simulator. The program runs at real-time speed. Processing of all obstacles (i.e., construction of the polygonal representations, computation of the force histograms and generation of the linguistic descriptions) is done faster than the robot can move, so there are no delayed results.

6   Conclusion

We have shown in this chapter that the notion of the histogram of forces can be of great use in understanding the spatial organization of image objects. This is a crucial problem, essential to countless domains of computer vision. The histogram of forces provides a fuzzy qualitative representation of the relative position between 2D objects. Because it offers solid theoretical guarantees and has nice geometric properties, it can be used in scene matching, and enables the pose parameters to be retrieved. It can also be exploited in pattern recognition. For instance, the F-signature—a particular force histogram that represents the shape of an object—has been used to classify cranium sinuses. Spatial databases can clearly benefit from a tool like the histogram of forces. We are currently working on new spatial indexing mechanisms for medical image databases. They rely on the computation of force histograms for modeling the relationships between lesions and anatomical landmarks. Geographic information systems constitute a promising ground also, especially as force histograms are able to handle vector data in a very efficient manner. Moreover, the histogram of forces lends itself, with great flexibility, to the definition of numerous fuzzy spatial relations. In particular, new families of fuzzy directional relations have been introduced. They preserve important relative position properties, and can provide inputs to systems for linguistic scene description. One such system has been developed and dedicated to human-robot communication. Using readings from a ring of sonar sensors, a mobile robot describes its spatial relationship with the environment. The program runs at real-time speed. In the future, we plan to address two important challenges. The first one is very interesting from a purely theoretical point of view (probably less from a practical point of view, although it might be useful in scene matching for instance). It consists in solving the inverse problem, i.e., finding all the pairs of objects associated with a given force histogram. The second challenge consists in extending the notion of the histogram of forces so that three-dimensional entities can be handled. Analysis of 3D magnetic resonance images and design of virtual environments are two examples of potential applications. We also plan to define new fuzzy spatial relations, especially new models of "surrounds," and to develop mechanisms to adapt fuzzy relations and spatial descriptions to individual users.

Acknowledgments

This work was supported in part by grant N00014-96-0439 from the Office of Naval Research. It is based on previously-published joint research with George Chronis, Jim Keller, Jonathon Marjamaa, Chi-Ren Shyu, Ozy Sjahputera, Marge Skubic, and Laurent Wendling.

References

1. I. Bloch. Fuzzy Relative Position between Objects in Images: a Morphological Approach. In *ICIP'96*, 2:987-990, Lausanne, 1996.
2. I. Bloch. Fuzzy Relative Position between Objects in Image Processing: New Definition and Properties Based on a Morphological Approach. *Int. J. of Uncertainty Fuzziness and Knowledge-Based Systems*, 7(2):99-133, 1999.
3. I. Bloch, H. Maitre, and M. Anvari. Fuzzy adjacency between image objects. *Int. Journal of Uncertainty Fuzziness & Knowledge-Based Systems*, 5(6):615-653, 1997.
4. D. Dubois and H. Prade. *Fuzzy Sets and Systems: Theory and Applications*. Academic Press, New York, 1980.
5. J. Freeman. The Modeling of Spatial Relations. *Computer Graphics and Image Processing*, 4:156-171, 1975.
6. P. D. Gader. Fuzzy Spatial Relations Based on Fuzzy Morphology. In *FUZZ-IEEE 1997 (IEEE Int. Conf. on Fuzzy Systems)*, 2:1179-1183, Barcelona, Spain, 1997.
7. K. P. Gapp. Basic Meanings of Spatial Relations: Computation and Evaluation in 3D Space. In *AAAI'94*, pages 1393-1398, Seattle, WA, 1994.
8. K. P. Gapp. Angle, Distance, Shape, and their Relationship to Projective Relations. In *17th Conf. of the Cognitive Science Society*, 1995.
9. J. M. Keller and L. Sztandera. Spatial Relations among Fuzzy Subsets of an Image. In *1st Int. Symposium on Uncertainty Modeling and Analysis*, pages 207-211, College Park, University of Maryland, 1990.
10. J. M. Keller and X. Wang. Comparison of Spatial Relation Definitions in Computer Vision. In *ISUMA-NAFIPS'95*, pages 679-684, College Park MD, 1995.
11. J. M. Keller and X. Wang. Learning Spatial Relationships in Computer Vision. In *FUZZ-IEEE 1996*, 1:118-124, New Orleans, 1996.
12. G. Klir and T. Folger. *Fuzzy Sets, Uncertainty, and Information*. Englewood Cliffs, NJ: Prentice Hall, 1988.
13. L. T. Kòczy. On the Description of Relative Position of Fuzzy Patterns. *Pattern Recognition Letters*, 8:21-28, 1988.

14. R. Krishnapuram, J. M. Keller, and Y. Ma. Quantitative Analysis of Properties and Spatial Relations of Fuzzy Image Regions. *IEEE Trans. on Fuzzy Systems*, 1(3):222-233, 1993.

15. J. Marjamaa, O. Sjahputera, J. Keller, and P. Matsakis. Fuzzy Scene Matching in LADAR Imagery. In *FUZZ-IEEE 2001 (IEEE Int. Conf. on Fuzzy Systems)*, Melbourne, Australia, December 2001.

16. P. Matsakis. *Relations spatiales structurelles et interprétation d'images*. Ph. D. Thesis, Institut de Recherche en Informatique de Toulouse, France, 1998.

17. P. Matsakis and S. Andréfouët. The Fuzzy Line Between Among and Surround. In *FUZZ-IEEE 2002 (IEEE Int. Conf. on Fuzzy Systems)*, Honolulu, Hawaii, May 2002, to appear.

18. P. Matsakis, J. Keller, O. Sjahputera, and J. Marjamaa. Image Object Pair Matching with Camera Pose Estimation. *PAMI (IEEE Pattern Analysis and Machine Intelligence)*, submitted.

19. P. Matsakis, J. Keller, and L. Wendling. F-histogrammes et relations spatiales directionnelles floues. In *LFA'99 (French-Speaking Conf. on Fuzzy Logic and Its Applications)*, 1:207-213, Valenciennes, France, October 1999.

20. P. Matsakis, J. Keller, L. Wendling, J. Marjamaa, and O. Sjahputera. Linguistic Description of Relative Positions in Images. *TSMC Part B (IEEE Trans. on Systems, Man and Cybernetics)*, 31(4):573-588, 2001.

21. P. Matsakis and L. Wendling. A New Way to Represent the Relative Position between Areal Objects. *PAMI (IEEE Trans. on Pattern Analysis and Machine Intelligence)*, 21(7):634-643, 1999.

22. P. Matsakis and L. Wendling. Classification d'orbites et de sinus s'appuyant sur le calcul d'histogrammes de forces. In *RFIA'2000 (French-Speaking Conf. on Pattern Recognition and Artificial Intelligence)*, 1:111-118, Paris, France, February 2000.

23. P. Matsakis and L. Wendling. Orbit and Sinus Classification based on Force Histogram Computation. In *ICPR'2000 (15th Int. Conf. on Pattern Recognition)*, 2:451-454, Barcelona, Spain, September 2000.

24. P. Matsakis, L. Wendling, and J. Desachy. Représentation de la position relative d'objets 2D au moyen d'un histogramme de forces. *Traitement du Signal*, 15(1):25-38, 1998.

25. K. Miyajima and A. Ralescu. Spatial Organization in 2D Segmented Images: Representation and Recognition of Primitive Spatial Relations. *Fuzzy Sets and Systems*, 65(2/3):225-236, 1994.

26. K. Miyajima and A. Ralescu. Spatial Organization in 2D Segmented Images. In *FUZZ-IEEE 1994 (IEEE Int. Conf. on Fuzzy Systems)*, pages 100-105, Orlando, FL, 1994.

27. G. Retz-Schmidt. Various Views on Spatial Prepositions. *AI Magazine*, 9(2):95-105, 1988.

28. A. Rosenfeld. Fuzzy geometry: an updated overview. *Information Sciences*, 110(3/4): 127-33, 1998.

29. A. Rosenfeld and A. C. Kak. *Digital Picture Processing*, Academic Press, 2:263-64, 1982.

30. A. Rosenfeld and R. Klette. Degree of Adjacency or Surroundedness. *Pattern Recognition*, 18(2):167-177, 1985.

31. C. Shyu and P. Matsakis. Spatial Lesion Indexing for Medical Image Databases Using Force Histograms. In *CVPR'2001 (IEEE Int. Conf. on Computer Vision and Pattern Recognition)*, Hawaii, December 2001.

32. M. Skubic, G. Chronis, P. Matsakis, and J. Keller. Spatial Relations for Tactical Robot Navigation. In *AeroSense 2001 (SPIE Int. Symposium on Aerospace/Defense Sensing, Simulation, and Controls)*, vol. 4364, Orlando, Florida, USA, April 2001.

33. M. Skubic, P. Matsakis, G. Chronis, and J. Keller. Generating Multi-level Linguistic Spatial Descriptions from Range Sensor Readings Using the Histogram of Forces. *Autonomous Robots*, submitted.

34. M. Skubic, P. Matsakis, B. Forrester, and G. Chronis. Extracting Navigation States from a Hand-Drawn Map. In *ICRA'2001 (IEEE Int. Conf. on Robotics and Automation)*, pages 259-264, Seoul, Korea, May 2001.
35. X. Wang and J. M. Keller. Fuzzy surroundedness. In *FUZZ-IEEE 1997 (IEEE Int. Conf. on Fuzzy Systems)*, 2:1173-1178, Barcelona, Spain, 1997.
36. L. Wendling, P. Matsakis, and S. Tabbone. Fast and Robust Recognition of Orbit and Sinus Drawings Using Histograms of Forces. *Pattern Recognition Letters*, under revision.
37. S. W. Zucker. Region Growing: Childhood and Adolescence. *Computer Graphics and Image Processing*, 5:382-399, 1976.

Fuzzy Spatial Relationships and Mobile Agent Technology in Geospatial Information Systems

Frederick E. Petry [1] [*], Maria A. Cobb [2], Dia Ali [2], Rafal Angryk [3],
Marcin Paprzycki [2], Shahram Rahimi [3], Lixiong Wen [4], Huiqing Yang [3]

[1] Naval Research Laboratory
Mapping, Charting & Geodesy
Stennis Space Center, MS 39529
petry@eecs.tulane.edu

[2] Department of Computer Science & Statistics
University of Southern Mississippi
Hattiesburg, MS 39406-5106
firstname.lastname@usm.edu

[3] Center for Computational Science
University of Southern Mississippi
Hattiesburg, MS 39406-5106
firstname.lastname@usm.edu

[4] Department of Electrical Engineering &
Computer Science
Tulane University
New Orleans, LA 70118-5674

Abstract. This chapter discusses an integrated work in the definition and implementation of sets of fuzzy spatial relationships concerning topology and direction. We present our basic approach to defining these relationships as an extension to previous work in temporal relations. We also discuss several extensions to this approach that include refinements and alternate definitions. Two implementations are also described, one in a C++, Oracle database environment and another utilizing the expert system shell Fuzzy Clips. Finally we discuss the integration of this querying approach in an agent-based framework. Agent technology has become a leading implementation paradigm for distributed and complex systems, and has recently garnered much interest from researchers in the area of spatial databases. Agents offer many advantages with respect to intelligence abilities and mobility that can provide solutions for issues related to uncertainty in spatial data, such as those of spatial relationships.

1 Introduction

The need to handle imprecise and uncertain information concerning spatial data has been widely recognized in recent years, e.g., [19], particularly in the field of geographical information systems (GIS). GIS is a rather general term for a number of approaches to the management of cartographic and spatial information. Most definitions of a GIS [16,22] describe it as an organized collection of software systems and geographic data able to represent, store and provide access for all

---

[*] On sabbatical from Tulane University, New Orleans, LA, USA.

forms of geographically referenced information. At the heart of a GIS is a spatial database. The spatial information describes the location and shape of geographic features in terms of points, lines and areas.

There has been a strong demand to provide approaches that deal with inaccuracy and uncertainty in GIS. The issue of spatial database accuracy has been viewed as critical to the successful implementation and long-term viability of GIS technology [19]. There are a variety of aspects of potential errors in GIS encompassed by the general term "accuracy." However, here we are only interested in those aspects that lend themselves to modeling by fuzzy set techniques.

Many operations are applied to spatial data under the assumption that features, attributes and their relationships have been specified a priori in a precise and exact manner. However, inexactness often exists in the positions of features and the assignment of attribute values and may be introduced at various stages of data compilation and database development. Models of uncertainty have been proposed for spatial information that incorporate ideas from natural language processing, the value of information concept, non-monotonic logic and fuzzy set, evidential and probability theory. For example, in [32] there are reviews of four models of uncertainty based on probability theory, Shafer's theory of evidence, fuzzy set theory and non-monotonic logic. Each model is shown as appropriate for a different type of inexactness in spatial data. Inexactness is classified as arising primarily from three sources. "Randomness" may occur when an observation can assume a range of values. "Vagueness" may result from imprecision in taxonomic definitions. "Incompleteness of evidence" may occur when sampling has been applied, there are missing values, or surrogate variables have been employed. An excellent collection of recent papers on vague boundaries in spatial applications can be found in [4]. Various topics in the volume include some areas of particular interest such as topological relations and indeterminate boundaries, data models for indeterminate objects and fields, and vague shape models.

Robinson [29,28,27] has done early extensive research on fuzzy data models for geographic information. He has considered several models as appropriate for this situation—the two early fuzzy database approaches using simple membership values in relations [18,2], and a similarity-based approach [3]. In modeling a situation in which both the data and relationships are imprecise, he assesses that this situation entails imprecision intrinsic to natural language which is possibilistic in nature [37].

There have been several recent special sessions on fuzzy sets and GIS at FUZZ-IEEE'98, IPMU'98 and NAFIPS'99. A collection of such work has been collated into a special issue of *Fuzzy Sets and Systems* [12]. In the issue are found a number of approaches to the use of fuzzy sets and spatial data. Relevant selections include fuzzy objects for GIS [13], landform classification with fuzzy k-means [5], and fuzzy spatial queries [33].

In this chapter we describe the development of an approach for fuzzy querying using spatial relations, focusing especially on directional querying. We first provide an overview of the background work in this area, including, in particular, spatial relationships using Minimum Bounding Rectangles (MBRs) and fuzzy set-based spatial data approaches.

Next we describe our basic approach [11] using a variant of the spatial intervals and an underlying model called an abstract spatial graph (ASG) to support fuzzy querying. Following, we consider further development of the approach, such as the use of finer partitions know as MRRs to enhance querying. Then we discuss some modifications to resolve certain outlying cases that produce anomalous query results. Two implementations are also described, one in a C++, Oracle database environment and another utilizing the expert system shell, Fuzzy Clips. Finally we discuss the integration of this querying approach in an agent-based framework.

2 Background

Relevant background research includes various aspects of spatial reasoning, work on directional and topological relationship representation and the incorporation of uncertainty and fuzzy querying.

A basis for many researchers' approaches is the extension of Allen's temporal relations [1] to two or more dimensions for spatial reasoning. Examples of how this has been done can be found in [20,31,24,23] to name a few. For each of these, the approach taken is somewhat different, based on the intent of the work. However, the concept of representing a 2-D object as a set of two intervals, an x and a y, and of having the resulting spatial relationship consist of some combination of the component 1-D relations seems to underlie most.

Also relevant is [14], which presents a unified framework for approximate spatial and temporal reasoning using topological constraints as the representation schema and fuzzy logic for representing imprecision and uncertainty. The application of the resulting fuzzy representation to each of Allen's interval relationships is developed as the possibility of the occurrence of the conditions of the original definition. A different approach based on statistical methods for representing and deriving topological relations is given in [35]. The relations used are those in [15] restricted to 1-D, for which Winter introduces a partial ordering based on Galtons' ordering of topologic relations for space and time sequences [17]. The derivation of uncertain topologic relations is treated as a classification problem. As such, specific conditions for deriving relationship probabilities based on the testing of dominant vs. dominated relations are presented.

Additionally, all of this work appears to utilize the fact that Allen's interval relations and corresponding logic have relevance in two dimensions to two extremely significant areas in spatial reasoning: topology and direction (orientation). Thus, the various extensions to Allen's work have provided sound foundations from which to launch work in qualitative spatial reasoning. Sharma's work, in particular, takes advantage of the dual benefit of the model by showing how inferences can be made over the composition of topological and directional information [31]. This extension includes a mapping of the temporal relations onto the eight mutually exclusive binary topological relations of the 4-intersection [15], a generic model that defines topological relationships through the intersections of boundaries and interiors of point sets.

The approach we have taken, as well as that of Nabil [24], Sharma [31] and Clementini [8], relies upon the use of MBRs as approximations of the geometry of spatial objects. An MBR is defined as the smallest X-Y parallel rectangle which completely encloses an object. The use of MBRs in geographic databases is widely practiced as an efficient way of locating and accessing objects in space. In addition, numerous spatial data structures and indexing techniques have been developed that exploit the computationally efficient representation of spatial objects through the use of MBRs [21,30]. Recently, Papadias and Theodoridis [25] have considered various indexing issues (R-trees, KDB trees etc.) to process topological and directional queries using MBRs. Their particular focus has been experimentation with alternative indexing to provide query optimization.

3   Fuzzy Directional Relationships and Querying

The core of the approach we shall describe is directly dependent upon the definitions of binary relationships between two-dimensional objects. For our purposes, we are assuming that images have been segmented and labeled, and that objects representing features have had MBRs assigned. We utilize an extension into the spatial domain of Allen's temporal relationships [1] where he defined a set of thirteen relationships that completely represents any relationship that can exist between two one-dimensional (temporal) intervals. These relationships are *before, equal, meets, overlaps, during, starts* and *finishes,* along with their inverses.

The spatial extension is as follows: given the MBRs of two objects, the binary relationship between the objects in both the horizontal and vertical directions can be completely defined by a tuple, $[r_x, r_y]$, where r_x is the one of Allen's temporal relations that defines the interaction of the object MBRs in the x direction, and r_y represents the same for the y direction. This results in a total of 85 possible relationships—49 base relations and 36 inverse relations.

The formal definitions for each of the relationships is given in terms of a set of constraints based on corner positions, each of which must hold between the MBRs for each object. For example, for the case of A [finishes,starts] B , the definition is given as: { $B_{xl} < A_{xl} < B_{X2}$, $A_{X2} = B_{X2}$, $B_{yl} < A_{y2} < B_{y2}$, $A_{yl} = B_{yl}$ }, where (xl,y1} and {x2, y2} represent the lower left and upper right corners, respectively, of the MBRs. In Figure 1 is an example set of object MBRs and a subset of the existing relationships between.   This set of MBRs and relationships will be used to demonstrate the modeling and query framework.

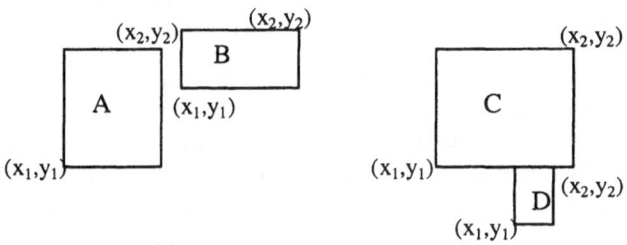

Fig. 1. MBR relationships.
A [before, overlaps] B; B [before, overlaps $^{-1}$] C; D [during, meets] C.

These spatial relationships can be used for qualitative topological relationship definitions. We   include in our set of "qualitative topological relationships" definitions that are intuitive and useful from a user's perspective, but which do not necessarily satisfy the property of topological invariance. Such definitions evolve by recognizing that logical groupings of the relationships can be used to define common topological concepts. The partitioning of the relationships in this manner can result in as many or as few groupings as desired. Such partitionings can be represented as sets, each of which is then associated with a linguistic term.

The implication of this is that any relationship that is contained within a particular set can be considered an independent representation of the corresponding qualitative topological relationship, represented by its linguistic term. Some examples are shown below using the notation of representing one of Allen's relations by its initial letter.

surrounded-by := { [dd] }

partially-surrounded-by := { [df] [fd] [do] [ds] [od] [sd] [do^{-l}] }

overlapped-by := { [oo^{-l}] [os^{-l}] [of^{-l}] }

Such definitions serve merely as examples, and may be redefined according to individual needs.

Next, the basic relationship definitions can be used in a similar manner for defining directional relationships. Specifically, the relationships with which we concern ourselves here are the cardinal directions and refinements: N, S, E, W, NE, SE, SW, NW. Given that the context for consideration is that of directional relationships between two-dimensional objects, it is apparent that a simple, crisp representation of direction is not adequate for supporting any but the simplest of query capabilities. Given the spatial extent of two-dimensional objects, it is very likely that in any one case, more than one of the eight directions listed above will apply, to either a greater or lesser degree. Therefore, we have developed a method for defining directional relationships that would allow for fuzzy querying of *any* of the directional relationships that exists between two objects.

The concept of *object sub-groups* is used as a basis for determining the set of directions that defines the directional relationship between two objects. Object sub-groups, similar in nature to ortho-relational objects described in [7], are partitions of the MBRs created by the extension of the edges of one or both of the MBRs involved in one of the 85 previously defined relationships. In those cases for which no extensions intersect the partner MBR, the entire object is considered to be an object sub-group. Additionally, any overlapping portion of the MBRs is defined as an object sub-group. Examples of MBR partitioning into object sub-groups for some example relationships are shown in Figure 2.

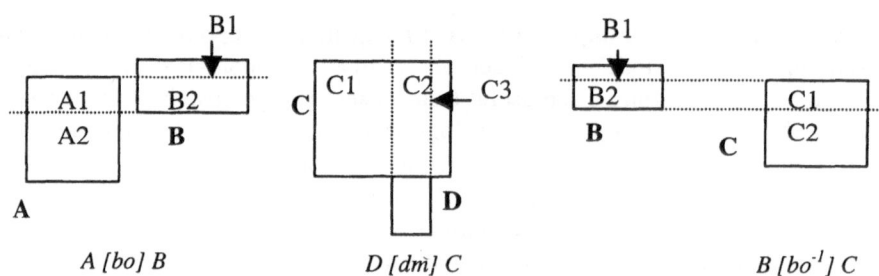

Fig. 2. Examples of object sub-group partitionings.

A *direction set is* then derived from this partitioning by identifying all possible directions that can be associated with any of the object sub-groups of A with respect to all of the object subgroups of object B. Given this, the direction sets corresponding to the object sub-groups shown in Figure 2 are:

{W, SW} for A [bo] B; {W, NW} for B [bo $^{-1}$] C; {S, SW, SE} for D [dm] C.

Direction sets were derived in this manner for each of the 85 relationships. The result was the creation of neighborhoods of sizes two and four for which the direction sets were equivalent for the constituent relationships. Direction sets for the relationships grouped by neighborhoods, which are considered as *equivalence classes* for directional relationships. All relationships within an equivalence class are represented by the corresponding direction set, and are considered equivalent with respect to direction. The creation of equivalence classes within and between the base and inverse relationship groups results in 27 possible direction sets.

Definitions for directions can now be defined in a manner analogous to the way in which qualitative topological relationships were defined earlier. The definition for any particular direction includes the set of all relationships containing that direction as a member of its direction set. The definition for the direction East is shown below as an example.

$$E ::= \{[dd], [df], [fd], [do], [ds], [ff], [d=], [fo], [fs],$$
$$[f=], [dd^{-1}], [do^{-1}], [ds^{-1}], [fd^{-1}], [df^{-1}], [fo^{-1}], [fs^{-1}]\}$$

The basic relationship definitions and their use in defining relevant directional and qualitative topological relationships provide a framework for the *abstract spatial graph* (ASG), a spatial data structure specifically designed to retain orientation and topological information with respect to two-dimensional objects, and to provide information to support fuzzy querying capabilities on these relationships.

The first step in constructing the ASGs is to categorize the original 85 relationships according to the level of interaction of the MBRs into four distinct categories: disjoint, tangent, overlapping and containment. Similar generalizations of relationships have been proposed, for example, in [23]. These categories provide sufficient distinction for formulating the definitions for the concepts of *reference areas* and *reference points* necessary for establishing an ASG for a particular binary relationship.

The general concept of a reference area is that of some region which is either common to two objects, or which can be intuitively derived from their relationship. A reference point is taken to be the centroid of the reference area. It is at the reference point that a directional axis is centered for the purpose of constructing the ASG. The four categories of relationships, along with definitions and examples for reference areas and reference points for each are shown in Figure 3. In the figure we see that for the disjoint relationship the reference area is the region between the two objects bounded on two sides by the neighboring sides of the MBRs and on the other two sides by appropriate extensions of the MBR sides (horizontally, from the leftmost object). The centroid of the reference area is the reference point in this case. The reference area, or line in the case of tangency, is the common tangent line segment and its center is the reference point. For overlapping figures, the reference area is the area of overlap, and for the case of the containment relationship, the reference area is the MBR of the contained object. In both of these cases the reference point is the centroid of the reference area.

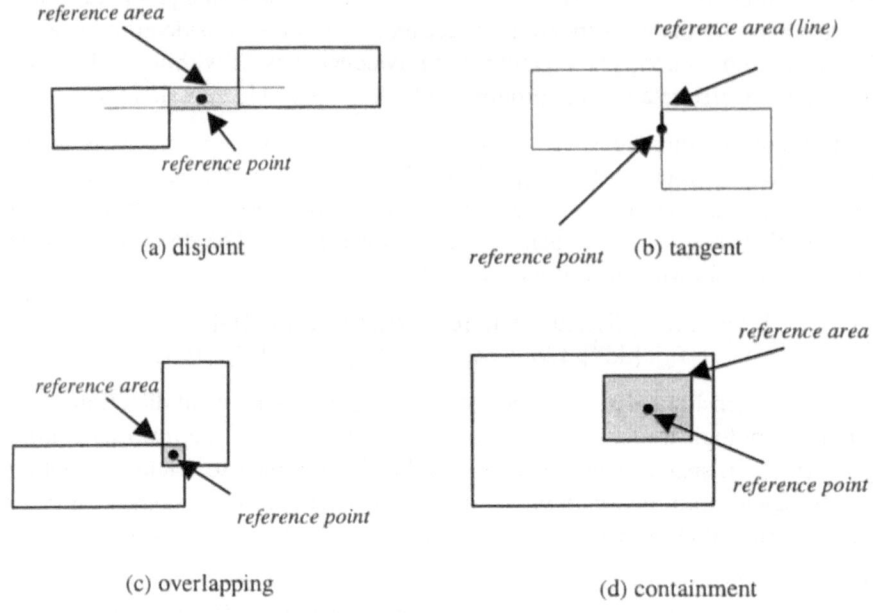

Fig. 3. Reference areas and reference points.

Again, the construction begins by centering a directional axis containing line segments for each of the eight compass points at the reference point of the relationship. Then, for each object sub-group through which an axis passes, a node for that object is placed on the graph at an orientation corresponding to that of the axis that crosses the object sub-group.

To avoid "cluttered" graphs that contain many nodes with low information content, it is desirable to set a minimum threshold on axis lengths (defined as the length of the axis segment from its entrance into an object sub-group to its exit from the same sub-group) such that for any situation in which the axis length is below the set threshold, no node is constructed. Such a threshold is obviously application dependent, and possibly even variable within an application.

Pictorially, nodes are each placed an equal distance from the origin. Different node representations (e.g., color-coded) are used for nodes representing object sub-groups belonging to different objects. An arc is then drawn to connect all of the nodes that represent the same object. The origin symbol differs depending upon the relationship's membership in one of the four basic categories.

In addition to providing information directly relevant to the representation of the abstract spatial graph, we also needed to represent ancillary information that can be used for fuzzy query inferences. This information is represented in the form of node "weights" that can be used for defining fuzzy topological and directional qualifiers for use with a fuzzy query framework.

Calculation of weights uses both the areas of object sub-groups and the lengths of axes that pass through object sub-groups. Three different types of weights are computed: *axis weights, area weights* and *node weights.* Each of these is derived in a specific manner designed to support a given objective for fuzzy querying.

Axis weights are an intermediate step for calculation of node weights. First, all axes whose lengths are less than some set minimum threshold are discarded, ensuring that we are dealing only with those axes whose lengths are significant. Then, the longest of the remaining axes is normalized to 1, and the weight of each of the other axes is computed as a ratio of its axis length to the length of the longest axis.

Area weights are also used in the calculation of node weights; however, these also have significance of their own. Area weights are calculated for each object sub-group and are defined as the ratio of area occupied by the object sub-group to the area of the entire object, as defined by the MBR. Finally, a total node weight is derived by multiplying the axis weight by the area weight for each node of the ASG.

Area weights and total node weights are stored for each node of the ASG, with the only exception being the origin node, which stores only an area weight. Since the origin node represents the reference area itself, axis length is not a reasonable consideration for the reference area object sub-group. Therefore, a node weight for the origin node can not be computed, and we allow the area weight to serve as the total node weight also.

The area weights and total node weights of ASGs directly support fuzzy queries regarding qualitative topological and directional information in two specific ways. Area weights provide an indication of the degree to which an object participates in a qualitative topological relationship. By mapping ranges of area weights to linguistic qualifiers such as *some, most,* etc., fuzzy information such as "some of object A overlaps most of object B," can be determined.

Total node weights, on the other hand, are used to indicate the extent to which one object can be considered to lie in a certain direction with respect to a second object. Again, ranges of weights can be correlated to linguistic terms, e.g., *slightly, mostly,* to provide qualifiers for directional orientation. Then, for example, one could determine that, while object A is *slightly southwest* of object B, it is at the same time *mostly west* of object B.

The preservation of all directional information regarding two objects, along with the use of total node weights, allows users to obtain a complete conceptualization of directional relationships between the objects. Furthermore, the calculation of the total node weight as the product of the axis weight and area weight ensures against bias in those cases, for example, where the object sub-group associated with a directional axis is very large (increasing the weight for that direction), but for which the axis weight is very small (indicating a weaker association for that sub-group/direction pair than for others for the same object).

By using the following such assignment for area weights:

{all(96 - 100 %), most(60 - 95 %), some(30 - 59 %), little(6 - 29 %), none(0 - 5 %)}

and node weights:

{directly(96-100 %), mostly(60-95 %), somewhat(30-59 %), slightly(6-29 %), not(0-5 %)}

we can determine for our example of Figure 1 that:

1. B is *mostly* west of C   3. D is *directly* south of C

2. *Little* of B is northeast of A   4. C is *slightly* southeast of B

4   Extensions to the Model

4.1   Extensions to the Standard MBR Representation

The ASG model utilizes MBRs to represent features as described above, while algorithms for computing relationships for the model are designed in such a way as to minimize the potentially negative impact of the use of MBRs. Additionally, the ASG model extends the use of rectangular boundaries as representations of sub-objects within the MBRs.

In keeping with generally accepted practice, both MBRs and the sub-rectangles utilized in the ASG model retain x-y-parallelism; however, in this section we explore several variations on this traditional approach and investigate the respective implications to the modeling of relationships. Our goal in investigating alternative representations for geometric properties of spatial features is threefold: (1) alleviate or significantly decrease anomalies in topological relationship determination based on MBRs; (2) improve accuracy in determination of directional and topological relationships between representations; and (3) maintain, as much as possible, computational efficiency in processing concerning relationship categorization and querying.

First, we consider the implications of partitioning MBRs into sets of rectangles, essentially resulting in a gridded surface which is an approximation we call *Multiple Rectangle Representation,* or MRR.   Three variations on MRRs which we analyze in this section include (1) a uniform MRR, (2) a non-uniform, congruent MRR, and (3) a non-uniform, non-congruent (irregular) MRR.   All three variations of MRRs result in a finer approximation of the object's true geometry than do MBRs, while maintaining a basic regular, rectangular structure for which computationally efficient methodologies have been developed.

The first variation can be viewed as the imposition of a grid of any level of resolution upon an object, after which any of the rectangles not actually intersecting with a part of the object is discarded. Two cases include one in which grids of the same resolution are used for both objects participating in a relationship, as well as the case in which grids of different resolutions are used. Figure 4 shows a simple example of the use of uniform MRRs. The dotted line shows the original boundaries.

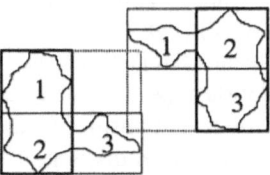

Fig. 4. Uniform MRRs used for object representations.

Now we consider enhancements to the ASG to accommodate the use of MRRs for relationship determination. We begin with the supposition that the set of ASGs is a closed set, such that any modification to the way in which relationships are defined does not result in any new ASG. It is apparent, however, that the way in which the ASGs themselves are defined for MRRs must be modified to take advantage of the more accurate representation.

We do this by first computing a set of ASGs—one ASG for every combination of relationships between sub-rectangles of both objects' MRRs. This results in a set of ASGs for each binary relationship, $S = \{A_{ij} \mid 1 <= i <= n, 1 <= j <= m$, where n = number of subrectangles in 1st object, m = number of subrectangles in 2nd object}. For example, in Figure 4 are two objects with simple, uniform MRRs. The resulting set of relationship ASGs for this example is:

$$S = \{A_{11} = [bo], A_{12} = [bo], A_{13} = [bo^{-1}], A_{21} = [bb],$$
$$A_{22} = [bb], A_{23} = [bo], A_{31} = [ob], A_{32} = [bb], A_{33} = [bo]\}.$$

An approach for utilizing this information set to gain a more accurate picture of the relationship under consideration is to first associate a count with every distinct relationship appearing in the set. For our example, this would yield $S' = \{([bo], 4), ([bb], 3), [bo^{-1}], 1), ([ob], 1)\}$. These counts divided by the total number of sub-rectangle relationship combinations (9 in the example) are considered as membership values in fuzzy relationships. Rather than having a single, crisp binary relationship as was the case in the original ASG model, we now have a *set* of ASG relationships, each member of which a given binary relationship belongs to some degree. For example, the fact that [bo] and [bb] appear as the predominant relationships for Figure 4, along with the fact that [bo^{-1}] and [ob] exist, although to a much lesser degree, conveys a substantially more significant amount of information than does the original [oo] designation, which in this case is also inaccurate with respect to the contained objects' relationship due to the topological inconsistency problem associated with MBRs.

We then take this a step further by associating these membership values with the higher-level relationships defined in the previous section [10]. Because these relationships represent a mutually exclusive, total partitioning of the basic relationships, a mapping from the members of S to these relationships will result in at most the same number of relationships as the number of members of S. However, it is more often the case that fewer high-level relationships result. This is because (1) such relationship definition sets are often composed of basic relationship neighbors, and (2) the use of MRRs in the manner described necessarily implies relationship categorizations for neighboring sub-rectangles, resulting in neighboring relationships. In our example, each of [bo], [bb], [bo^{-1}], and [ob] appears in the *disjoint* relationship, leading to the invariable conclusion that the two representations are indeed disjoint. This approach eliminates the need to compute the set of weights for ASG nodes as was done in the original model, as similar information is now maintained in S′.

The application of non-uniform, congruent MRRs can be understood as an analog to a quadtree decomposition commonly performed for spatial indexing purposes. The approach begins with standard x-y-parallel MBRs, upon which a quadtree-like decomposition is performed, with the MBR being divided into four equivalent rectangles, any or all of which can then be divided similarly, continuing until as fine a partitioning as desired is achieved. At that point, any rectangles not actually containing a part of the object are discarded. We say the rectangle sets are *regularly hierarchical* because the level of detail (size) of the rectangles can vary across different parts of the object so as to achieve a desired level of representation, while each larger rectangle is exactly a multiple to the fourth of any of the smaller rectangles of the object. An example of this type of MRR is shown by the grayed area in Figure 5.

This type of boundary approximation more accurately represents the objects' geometry in comparison to either MBR or uniform MRR representations. While maintaining the greatly reduced incidence of topological inconsistencies between true and approximate boundaries illustrated in the uniform MRRs, additional levels of detail have been added that allow for more accurate relationship determination. Since the areas are still rectangular decompositions, computational issues remain simplified compared to boundary representations.

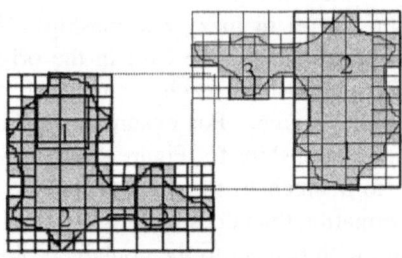

Fig. 5. A non-uniform congruent MRR.

The most significant issue that must be addressed concerning fuzzy relationship modeling for this approach is the manner in which the differently sized rectangles are assessed as contributors to overall relationship determination. This can be handled by extending the approach used for uniform MRRs. While for uniform MRRs it was sufficient to use the mere existence of the basic relationships between sub-rectangles as equal factors in fuzzy relationship categorizations, we must now compute a weight analogous to the ASG node weights for each sub-rectangle relationship to achieve a level of normalization for "combining" the individual relationships into one.

Recall that the area weights are computed as the ratio of the MBR sub-object areas to the total MBR area, and that these weights are used to identify the extent to which an object participates in a given relationship. Using this same approach for non-uniform congruent MRRs—calculating a weight as the ratio of a sub-rectangle's area to the combined area of all sub-rectangles containing a part of the object—we achieve a level of normalization for use of differently sized rectangles.

For each applicable relationship, the weights for every different sub-rectangle of each object for that relationship are summed, resulting in a value ≤ 1. Whenever one sub-rectangle participates in the same relationship with multiple sub-rectangles of the second object, that sub-rectangle's weight is only counted once. The resulting sums for a relationship for the two objects are then multiplied, yielding a weight for that relationship. For example, consider the case illustrated in Figure 6.

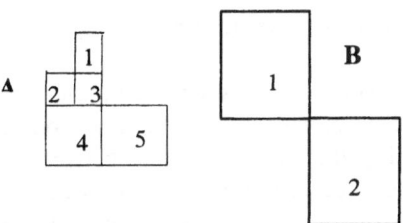

$A1_w = .1; A2_w = .1; A3_w = .1; A4_w = .35; A5_w = .35; B1_w = .5; B2_w = .5$

Fig. 6. Non-congruent MRRs with area weights.

For this example, we have the following set of relationships:

A \ B	1	2
1	[bd]	[bb^{-1}]
2	[bd]	[bb^{-1}]
3	[bd]	[bb^{-1}]
4	[bo]	[bo^{-1}]
5	[bo]	[bo^{-1}]

By summing and multiplying the area weights in the manner described earlier, we obtain the following :

$$[bd]_w = .15; [bo]_w = .35; [bb^{-1}]_w = .15; [bo^{-1}]_w = .35.$$

This shows that the relationships [bo] and [bo^{-1}] are weighted more heavily, primarily due to the larger areas of sub-rectangles 4 and 5 of object A. These weights can then be associated with the higher level relationships in the same manner as was shown for the uniform MRRs.

4.2 Geometric Modeling Capabilities

The use of MRRs for the ASG model has several implications. First is that the partitioning of MBRs in the ways described has *no effect* on the basic relationship definitions. Any relationship that originally held between two MBRs remains valid for the resulting MRRs, because minima and maxima for the objects do not change; therefore, any partitioning of the MBRs, regardless of granularity, will not affect the basic relationship between the geometric representations of the objects.

To support this statement, we begin by observing that the abstract spatial graphs can be arranged in a matrix according to the concept of *conceptual neighborhoods* with respect to the relationships that are represented. That is, each of the relationships that borders a particular relationship in both the horizontal and vertical directions can be derived from the given relationship without transitioning through any other relationship state. If a transition to an immediate neighboring relationship is impossible, then it follows that a transition to any other relationship is also impossible. We illustrate this by first examining the example shown in Figure 7, which represents a portion of the complete relationship matrix.

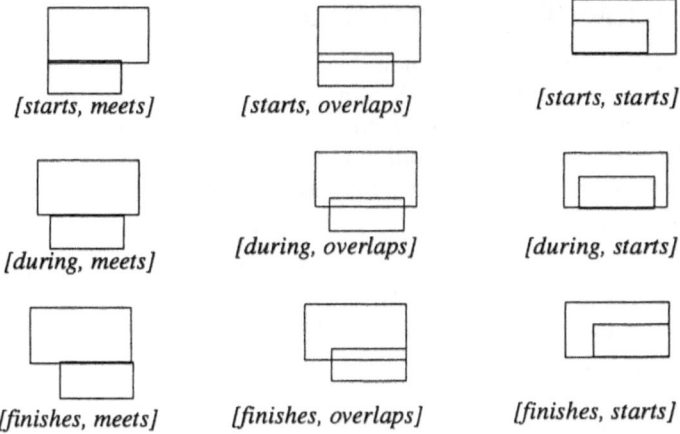

[starts, meets]	*[starts, overlaps]*	*[starts, starts]*
[during, meets]	*[during, overlaps]*	*[during, starts]*
[finishes, meets]	*[finishes, overlaps]*	*[finishes, starts]*

Fig. 7. [during, overlaps] relationship and conceptual neighbors.

Now we show that the enhanced accuracy of the boundaries for the MRR approximation model provides additional information in the way of geometric refinements for the relationship definitions. While there originally was only one geometric model for each relationship, there is now an infinite *set* of such models that correlates to any given relationship. Simple examples for a standard MRR approach are shown in Figure 8. Internal rectangle boundaries are omitted for clarity.

From this illustration, one can see that any selected MRR partitioning for a pair of MBRs will never cause a transition to one of the neighboring relationships. Therefore, we are able to operate under the assumption of a closed set of relationships for which the ASG model and query framework hold.

Figure 8 also demonstrates the advantages associated with the use of MRRs over MBRs: (1) the ability to better represent correct topology, and (2) the ability to make finer distinctions between geometric relationships. The first of these is illustrated in Figure 8(a) which shows that the two objects could not possibly overlap, while the original MBR representation shows an overlap. The second advantage can be seen in Figures 8(a)(b) and 8(c)(d) which show the same topological relationships—disjoint and overlaps—but for which geometric distinctions exist which provide additional information for spatial analysis.

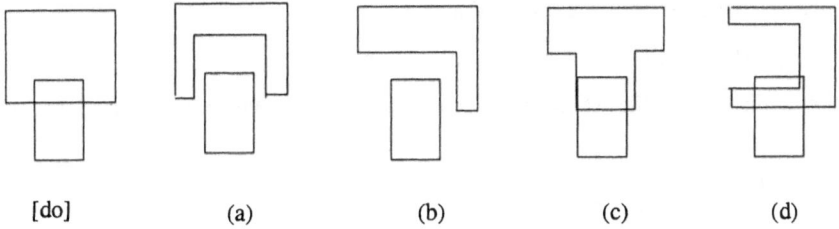

[do]   (a)   (b)   (c)   (d)

Fig. 8. A set of geometric representations for the [do] relationship.

4.3   An Extension for Expert System Implementation

Here we describe an implementation based on the C Language Integrated Production System (CLIPS). CLIPS is an expert system tool that provides a complete environment for the construction of rule and/or object-based expert systems [9,36]. Because of its portability, extensibility, capabilities, and low-cost, CLIPS has received widespread acceptance throughout the government, industry and academia.

Three rule sets are used to represent the basic structure of this model. Recalling that each object is represented by an MBR, the smallest definition of which is two diagonal points of the box, we consider two objects:

$A\ (A_{x1}, A_{y1})\ (A_{x2}, A_{y2})$

$B\ (B_{x1}, B_{y1})\ (B_{x2}, B_{y2})$

Now, taking one point from each object, that is,

$(A1, A2) = (A_{x1}, A_{x2})$ or (A_{y1}, A_{y2})

$(B1, B2) = (B_{x1}, B_{x2})$ or (B_{y1}, B_{y2}),

we will present the implementation process, beginning with definitions of three essential rule sets.

Rule Set 1: Define a set of non-ambiguous relationships.

Considering one direction, the spatial relation between object A and object B can be defined as in Figure 9. This is a partial set of the rules needed in actual CLIPS format.

```
1. IF < A2<B1 >        THEN      < before >
   IF < B2<A1 >        THEN      < before⁻¹ >
3. IF < A1<B1<A2<B2 > THEN < overlap >
   IF < B1<A1<B2<A2 > THEN < overlap⁻¹ >
```

Fig. 9. CLIPS rules for four non-ambiguous relationships.

Simply, this rule set can be expressed as:

$r_x = (b, m, o, f, d, s, =, b', m', o', f', s')$.

In the y-direction, the same rules can be applied. Moreover, there are two additional rules that apply:

1. $A(r_x^{-1}, r_y)B = B(r_x, r_y^{-1})A$

2. $A(r_x^{-1}, r_y^{-1})B = B(r_x, r_y)A$

<u>Rule Set 2:</u> Define a set of topological relationships.

Based on the eighty-five basic relationships, the full topological relation set is:

T ={disjoint, tangent, surrounded-by, partially-surrounded, surrounded-by, partially-surrounds, overlapped-by, overlaps, x-subspace, y-subspace, y-subspaced-by}

Figure 10 shows a subset of the rules for topological relationships.

```
IF <dd> THEN   <A surrounded-by  B>

IF <oo'|os'|of'> THEN   <A overlapped-by B>
```

Fig. 10.   Topological relationship rule set example.

<u>Rule Set 3:</u> Define the set of directional relationships.

Figure 11 shows one of the rules for directions in CLIPS.

```
IF    <dd|df|fd|do|ds|ff|d=|fo|fs|
      f=|dd'|do'|ds'|fd'|df'|fo'|fs'>

THEN  < A East B >
```

Fig. 11.   Example CLIPS rule for direction.

Topological relations have been found to be useful for increasing the speed of spatial queries [34].   Therefore, let us analyze the geometric characteristics of topological relationships. Except for the disjoint relation, all other relations have a similar geometry; that is, the reference area is part of both objects involved. Thus, the original topological relation set can be reduced or reclassified to a binary topological set:

$$T \rightarrow T' = \{disjoint, connected\}$$

This new topological relation set will be used in the CLIPS implementation.

For convenience of implementation and further investigation, the ASG is modified by mapping topological relationships to 9 nodes for both objects. Similarly, each node has associated weights. However, the weight in a node may be null depending on the different topological relations. Since each object is associated with its 9 nodes, it is not necessary to keep information as to whether a node belongs to object A or object B in the implementation.   This provides a flexible structure for fuzzy querying.

4.4   A CLIPS Implementation

Now we illustrate how CLIPS can be used to implement the binary spatial relationships given earlier. Because of the amount of computation involved in implementation, we can take advantage of the *deffunction* construct that allows the addition of new functions without having to recompile and relink CLIPS. Several user-defined functions are written by using the CLIPS *deffunction* construct, which can be executed by CLIPS interpretively.

As a rule-based shell, CLIPS stores knowledge in rules, which are logic-based structures. In the implementation, the basic three rules are defined by using *defrule* constructs. They provide the basic spatial information such as, *Object A is disjoint from Object B,* or *Object A is West of Object B.* For fuzzy querying purposes, extra functions and rules are defined that will support fuzzy querying. The implementation is directly dependent upon the reduced topological relation set and modified ASG mentioned above.

The facts are the critical resources for the querying. All details for binary spatial relations are contained in *deftemplate* facts. The type of information stored in the database includes the positions of two objects, the reference object, non-ambiguous relations, and topological and directional relationships.   The corresponding data structures are declared by using *deftemplate* syntax.

To represent 2-D relations extended from Allen's relations, the *deffunction* construct in CLIPS is utilized. With this construct, a new function that implements Allen's relations in 1-D is defined directly in CLIPS. The rules that define a set of non-ambiguous relationships are built by a *defrule* construct.

The basic queries are based on the primary topological set and directional set. In this kind of querying, the degree to which one object lies in a particular direction with respect to a second object is not of concern. Figure 12 shows part of the CLIPS rule structures for directional relationships.

4.5   Fuzzy Querying of Binary Spatial Relationships

Based on the new topological relation set and modified ASG data structure, we defined three rules and four functions to support the processing of fuzzy queries. Query processing strategies are described as follows:

Step1. Find the reference area.
Fuzzy variable *weights* store all fuzzy query information. In order to get weights for each node in the ASG, a reference area must first be found. The reference area is also treated as an MBR object. We take two middle points among the four points in each direction as the reference object position. It can be represented as $R = (Rx_1, Ry_1) (Rx_2, Ry_2)$.

```
(defrule define-directional-relation
         (relationship (object1 ?A&A)
         (relations ?r) (object2 ?B&~A))
 =>
  (if (numberp (member$ ?r (create$ od of
                    sd sf dd df fd  ff =d =f
                    ob' om' oo' os' ...... )))
    then   (bind ?dr1 North))


                    . . . . .
  (loop-for-count (?count 1 8) do
      (bind ?dr (nth$ ?count (create$ ?dr1
              ?dr2 ?dr3  ...... ?dr7 ?dr8)))
      (if (numberp (member$ ?dr (create$
                    North East ...... West )))
        then
           (assert (directional-relationship
               (object1 ?A)(d-relation ?dr)
               (object2 ?B)) ) ) )    )
```

Fig. 12.   Defrule for directional relationship.

Step2. Calculate weights.
Based on the binary topological relations, a general method was developed for connected relations. For example, $NW\_area = (Rx_1 - Ox_1)(Oy_2 - Ry_2)$, where R represents the reference object, and O represents the one of two objects investigated. By adding some constraints, the general method for connected relations can also be applied to disjoint relations.

Given two objects and their reference object, the *weights* function maps the object sub-group into 9 nodes for each object, and calculates the area weights and node weights. The CLIPS program passes nine arguments to *weights* function, that is, one for object identifier, four for object position, and four for reference position. The function asserts area weights to the corresponding nodes for fuzzy querying.

Step3. Get qualifier to implement Fuzzy querying.
To provide support for fuzzy query processing, the fuzzy variable *weights* is assigned to the corresponding linguistic terms qualifier. The *fuzzyTq* function defines the topological qualifiers that represent the linguistic terms for area weight. Similarly the *fuzzyDq* function defines the directional qualifiers that represent the linguistic terms for node weight.

The fuzzy set for topological qualifiers is described below:

$$\{ \text{all} \ (0.96 - 1), \text{most} \ (0.6 - 0.95), \text{some} \ (0.3 - 0.59)$$
$$\text{little} \ (0.06 - 0.29), \text{none} \ (0 - 0.05 \ ) \ \}$$

and for directional qualifiers is:

$$\{ \text{directly} \ (0.96 - 1), \text{mostly} \ (0.6 - 0.95), \text{somewhat} \ (0.3 - 0.59),$$
$$\text{slightly} \ (0.06 - 0.29), \text{not} \ (0 - 0.05 \ ) \ \}.$$

The *fuzzy-query* rule in Figure 13 provides the fuzzy querying information by calling *fuzzyTq* and *fuzzyDq* functions.

```
(defrule fuzzy-query
 ?f3 <-(nodes (objectname ?A&A)
              (C ?C_area )
              (N ?N_area ?N_len)
               . . . . . .
              (NW ?NW_area ?NW_len) )
 =>
 (if (neq ?A B ) then (bind ?obj B) )
 (loop-for-count (?count 1 8) do
   (bind ?dir (nth$ ?count (create$ North
                          ... North_West)))
     (bind ?area_w (nth$ ?count (create$
                   ?N_area ?NE_area ?E_area
                   . . .?SE_area ?NW_area)))
     (bind ?node_w (nth$ ?count (create$
                   ?N_len ?NE_len ?E_len
                   . . .?W_len ?NW_len)))
     (bind ?tq (fuzzyTq ?A ?area_w
                        ?dir  ?obj))
     (bind ?dq (fuzzyDq ?A ?node_w
                        ?dir ?obj))
     (if (and (neq ?tq non) (neq ?dq non ))
     then
       (printout t "query information" crlf) ) ) )
```

Fig. 13. Fuzzy query rule.

Consider two objects and their respective corner coordinates:

object A (1 , 1) (5 , 3) and
object B (4 , 1)(8 , 7).

When the *define-2D-relation* rule is fired, calling AllenRelation (1 5 4 8) will return 'o', and the second calling of AllenRelation (1 3 1 7) will return 's.' Finally, the relation 'os' is added to the *spatial-relation* fact.

When the *define-topological-relation* rule is fired, the topological information 'Object A overlaps Object B' is displayed. When the *define-directional-relation* rule is fired, 'Object A is South Object B, Object A is South West of Object B, and Object A is West of Object B' are provided for directional relations. When the *reference* rule is fired, the reference object R(4, 1) (5, 3) is asserted into the fact database. While the *get-weight* rule is firing, area weights and node weights are assigned into 9 nodes for each object. Finally, the *fuzzy-query* rule fires, providing the following result:

> *Most of Object A is West of Object B*
> *Object A is mostly West of Object B*
> *Most of Object A is mostly West of Object B*

4.6 Modifications for Anomalous Cases

The original definition of the axis weight of an object sub-group is the ratio of its axis length to the longest axis length of the object sub-groups and it is used to specify how directly an object sub-group lies in a given direction. Under further consideration, the axis weight definition may not directly deal with directions but intuitively is a magnitude ratio between two object sub-groups. In other words, the physical significance of such axis weights regarding to directional information may be somewhat unclear in certain situations. Additionally, when there is no axis going through the MBR, there is no axis weight for this object and so no total node weights can be obtained for the subsequent querying.

Another specific case in which the axis weight definition may not clearly distinguish the directional difference between objects is shown in Figure 14. The reference areas are exactly the same in both Figures 14(a) and 14(b). There is only one object sub-group for each MBR and one and only one axis going through them, therefore the axis weights for both MBRs are 1. However, in Figure 14, object B extends itself markedly to the east. The directional relationships between A and B are apparently different in these two cases. In linguistic terms, in Figure 14(a), object B lies more directly northeast (mostly northeast) of object A, while in Figure 14(b), object B leans much more to the east of A (slightly northeast). The same axis weight values for these two cases, which result in the same total node weights, do not adequately distinguish these intuitively different directional relationships.

Another refinement we have studied is shown in Figure 15, in which objects 2 and 3 are symmetric to axis NE, with 2 on the eastern side, while 3 is to the north. If we calculate axis weights and total node weights for MBR 2 and 3 regarding to MBR 1, the same values will be obtained. To further distinguish the difference in such cases, a parameter termed as close_EW was developed, which specifies whether the node leans to east-west axis or to the north-south axis. It is defined as 1 if the node leans to the east-west axis and 0 if the node leans to the north-south

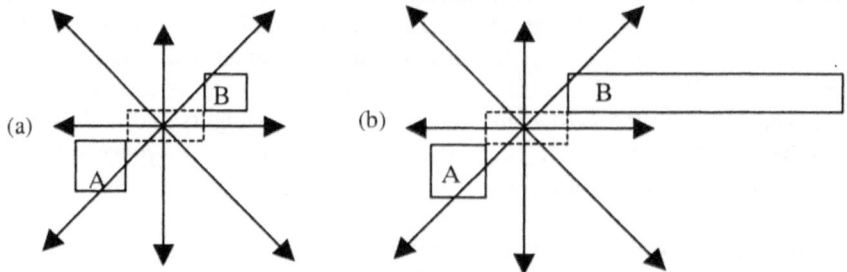

Fig. 14. Distinction between object directional relationships.

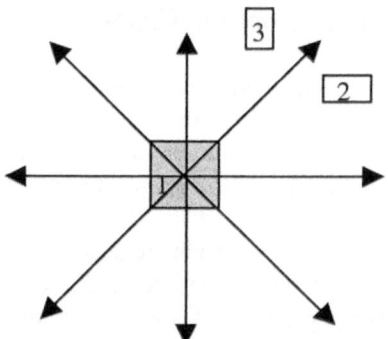

Fig. 15. Direction tendencies of objects.

axis. In Figure 15, close_EW of MBR 2 is 1 and that of MBR 3 is 0, which means that MBR 2 is leaning to the east, while 3 is to the north. In addition to specifying the direction tendencies, close_EW can also be used to further modify the total node weights for querying.

Total node weights can be used to query the directions of the object. Using a threshold on total node weights, as when the total node weight is below 5%, leads us to say the object is not in that direction. Again, somewhat non-intuitive situations can appear in these cases. In Figure 16, MBR 2 has only object sub-group 2 and its total node weight is below 5%. So MBR 2 is considered to be not northeast of MBR 1, but north of MBR 1 in this case. In fact, there is a range of such cases that appear to cause anomalies in an extreme limit situation.

Instead of dealing with all these cases separately when querying, we can formulate the different cases into generalized total node weights and query directions by such weights. From this point forward, the total node weights are intended to mean these generalized total weights, which are derived as follows:

Fig. 16. Object 2 may be considered directly north of object 1.

$(W_{TN})_i$: total node weight of object sub-group i (node i of ASG)

node i: 0 - overlap; 1 - north; 2 - northeast; 3 - east; 4 - southeast

5 - south; 6 - southwest; 7 - west; 8 - northwest

and

$(W_{TN})_0 = (W_{area})_0 * 1$

$(W_{TN})_1 = ((W_{area})_1 + (W_{area})_i) * 1$ i=2,8, if $(W_{node})_i < 0.05$ and $(close\_EW)_i = 0$

$(W_{TN})_i = (W_{node})_i$ i=2,4,6,8 and $(W_{node})_i >= 0.05$

or $(W_{TN})_i = 0$ i=2,4,6,8 and $(W_{node})_i < 0.05$

$(W_{TN})_3 = ((W_{area})_3 + (W_{area})_i) * 1$ i=2,4, if $(W_{node})_i < 0.05$ and $(close\_EW)_i = 1$

$(W_{TN})_5 = ((W_{area})_5 + (W_{area})_i) * 1$ i=4,6, if $(W_{node})_i < 0.05$ and $(close\_EW)_i = 0$

$(W_{TN})_7 = ((W_{area})_7 + (W_{area})_i) * 1$ i=6,8, if $(W_{node})_i < 0.05$ and $(close\_EW)_i = 1$

Now we need to consider only a single total node weight of the corresponding node to perform directional queries. To retrieve all objects that are directly north of the reference object, the querying becomes:

Obtain all objects with $(W_{TN})_1 >= 0.95$.

The total node weights are now generalized total node weights formed by summing up weights for all cases that lead to the same directional information. According to these modified total node weights, directional querying can be conducted by the definitions of directional relationships as shown in Figure 17.

	Directly	Mostly	Slightly	Somewhat	Not
Overlap	$(W_{TN})_0 >= 0.95$	$0.6<=(W_{TN})_0$ <0.95	$0.3<=(W_{TN})_0$ <0.6	$0.05<=(W_{TN})_0$ <0.3	$(W_{TN})_0 <0.05$
North	$(W_{TN})_1 >= 0.95$	$0.6<=(W_{TN})_1$ <0.95	$0.3<=(W_{TN})_1$ <0.6	$0.05<=(W_{TN})_1$ <0.3	$(W_{TN})_1 <0.05$
Northeast	$(W_{TN})_2 >= 0.95$	$0.6<=(W_{TN})_2$ <0.95	$0.3<=(W_{TN})_2$ <0.6	$0.05<=(W_{TN})_2$ <0.3	$(W_{TN})_2 <0.05$
East	$(W_{TN})_3 >= 0.95$	$0.6<=(W_{TN})_3$ <0.95	$0.3<=(W_{TN})_3$ <0.6	$0.05<=(W_{TN})_3$ <0.3	$(W_{TN})_3 <0.05$
Southeast	$(W_{TN})_4 >= 0.95$	$0.6<=(W_{TN})_4$ <0.95	$0.3<=(W_{TN})_4$ <0.6	$0.05<=(W_{TN})_4$ <0.3	$(W_{TN})_4 <0.05$
South	$(W_{TN})_5 >= 0.95$	$0.6<=(W_{TN})_5$ <0.95	$0.3<=(W_{TN})_5$ <0.6	$0.05<=(W_{TN})_5$ <0.3	$(W_{TN})_5 <0.05$
Southwest	$(W_{TN})_6 >= 0.95$	$0.6<=(W_{TN})_6$ <0.95	$0.3<=(W_{TN})_6$ <0.6	$0.05<=(W_{TN})_6$ <0.3	$(W_{TN})_6 <0.05$
West	$(W_{TN})_7 >= 0.95$	$0.6<=(W_{TN})_7$ <0.95	$0.3<=(W_{TN})_7$ <0.6	$0.05<=(W_{TN})_7$ <0.3	$(W_{TN})_7 <0.05$
Northwest	$(W_{TN})_8 >= 0.95$	$0.6<=(W_{TN})_8$ <0.95	$0.3<=(W_{TN})_8$ <0.6	$0.05<=(W_{TN})_8$ <0.3	$(W_{TN})_8 <0.05$

Fig. 17. Modified directional relationship definitions.

4.7 Oracle Implementation

An implementation was performed for the spatial direction approach with Visual C++ 6.0 and Oracle 8i. All calculations, displaying and interfacing were in Visual C++ 6.0. The Oracle 8i database was used to store the coordinates and the ASG weights of each MBR. That database was accessed from Visual C++ by using MFC's ODBC (Open Database Connectivity) database classes. Therefore, the MBR coordinates stored in the database can be obtained to calculate the weights and to display the MBRs, and the calculated weights can be sent to the database for subsequent queries from the programs.

A key issue for the implementation was how to store the total node weights, which are the basis for the directional querying. When directional relationships between two objects are of interest, one of the objects has to be the reference object and if it changes, the directional relationships of the object changes, as well as the total node weights for this object. When there are multiple objects, the issue must be resolved regarding which object to use as the reference for calculating the weights for all the other objects. These weights are then stored in the database. Another option is to pre-calculate each object's total node weights using every other object as a reference. This results in a set of total node weights being stored for each object, rather than a single weight. In this way, the total node weights between any combination of objects are calculated, eliminating the need for directional transitivity computations when querying. However, a huge amount of storage is required for this method, thus making it largely impractical for all but the smallest of datasets.

To resolve this problem, a dynamic database was utilized. A static table stores the coordinates of each MBR, and another table is used to store the total node weights for those MBRs dynamically. Initially, the total node weight table stores no information. Every time a query needs to be conducted, the reference MBR is set first and the total node weights are calculated accordingly and stored in the table, and the query is conducted upon this table. Any subsequent query based on the same reference MBR can be performed on this table without recalculation of the weights. If the reference MBR for the subsequent query changes, the stored information in the total node weight table is removed first, then the reference MBR for computation is reset, followed by recalculation and storage of the total node weights for all MBRs and the querying. By calculating and storing the total node weights dynamically, no directional transitivity needs to be considered, and the amount of information that needs to be stored is small. This method has also proven to provide satisfactory querying.

Directional Querying

Two types of directional querying were implemented in this work. The first one queries the directional relationships between any two MBRs (one of them is the reference MBR), such as "Get the directions of MBR A to MBR B (B is the reference MBR)." The other type of querying obtains all MBRs that lie in a specific direction of the reference MBR, for example, "Retrieve all objects that are directly northwest of object A."

When querying is on the directional relationships between two MBRs, only the total node weights of the queried MBR are calculated. The weights are not stored into the total node weight table of the database, but translated into linguistic terms like "directly north," "mostly southwest" or "somewhat south," and displayed to the user.

When the querying is to retrieve all MBRs in a specific direction from the reference MBR, total node weights for all MBRs except the reference MBR must be calculated. The weights are then stored into the total node weight table of the database and can be accessed by using the MFC's ODBC CRecordset class, DirectionWeightsDB. The querying is then realized by executing some SQL statements on the DirectionWeightsDB object.

We will now describe the first option as an example—retrieve the directional relationships between two objects. At first, the working MBR, for which directions are to be queried, must be set. Any existing MBR number can be entered in the dialog in Figure 18. In case of an invalid value, there is a prompt to re-enter a valid number. After setting the working MBR, its directions upon the reference MBR can be obtained by clicking on the "Get Directions" button and the directions will be displayed in the dialog box. For example, the working MBR in Figure 18 is MBR 3; its directions to MBR 0 are:

"Mostly east, somewhat north, somewhat south, and somewhat overlapping."

In addition, if the "Get Picture" is clicked, the working MBR will be re-displayed in red. Therefore, the two MBRs, the working MBR and the reference MBR, will stand out from the others for ease of evident visualization of their directional relationships.

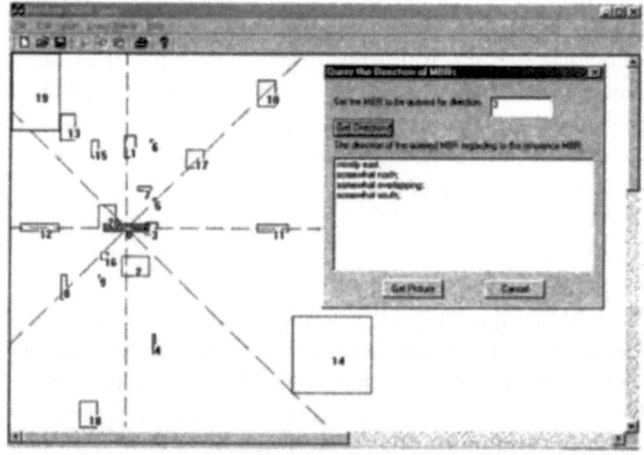

Fig. 18. User interface querying example.

5   Intelligent Agent Technology

In this section, we provide a brief introduction to the topic of intelligent agents, and present an agent-based framework for implementing the ASG model within a larger, distributed system.

5.1   Overview

We can define an agent as anyone or anything that acts as a representative for another party, for the express purpose of performing specific acts that are seen to be beneficial to the represented party. A software agent, which has been around for approximately a decade, is a software program that performs tasks for its user within a computing environment. Technically speaking, most fourth generation application software could be defined as agents. Every day we ask computers, through software, to perform hundreds of different tasks for us, essentially calling upon their agency attributes.

As we descend deeper into the concept of agency, we can see that there are distinct characteristics that collectively constitute an intelligent software agent. Intelligent software agents, or intelligent agents for short, are differentiated from other applications by their added dimensions of mobility, autonomy, and the ability to interact independent of its user's presence. When we introduce the additional element of intelligence to an agent, we must include the ability for adaptive reasoning. This implies the capability to process information from external environments — such as networks, databases, and the Internet — given a set of knowledge, attitudes, and beliefs of the user as understood by the agent.

There are six key fundamental characteristics of intelligent agents that differentiate them from other types of software applications [26]:

- Autonomy,
- Communication Ability,
- Capacity for Cooperation,
- Adaptive Behavior,
- Trustworthiness, and
- Capacity for Reasoning and Learning.

The ability to perform reasoning and learning is one of the key aspects of intelligence that distinguishes intelligent agents from other more "robotic" agents. As Belgrave [6] describes, reasoning implies that "an agent can possess the ability to infer and extrapolate based on current knowledge and experiences - in a rational, reproducible way." Based on our investigation there are five types of reasoning and learning scenarios:

- Rule-based reasoning, where agents use a set of user pre-conditions to evaluate conditions in the external environment,

- Knowledge-based reasoning, where agents are provided large sets of data about prior scenarios and resulting actions, from which they deduce their future moves,

- Simple Statistical Analysis for learning and reasoning,

- Fuzzy Agents for reasoning when the information is imprecise or incomplete,

- Neural Networks reasoning for unstructured data or noisy data, and

- Evolutionary Computing to expand the learning by viewing the system from an inter-agent perspective and employing a genetic algorithm.

5.2 Rule-Based Reasoning

Of all the technologies used to build intelligent agents, the easiest to understand is rule-based reasoning, the basis for "inference engines." Agents use the set of rules to decide which action or actions they should take ("If a condition C is satisfied, then perform action A"). With multiple rules, one rule's action may cause the satisfaction of another rule's conditions. This kind of chained effect is called forward chaining, and the problem with this approach is that the user needs to recognize the opportunity for employing an agent, take the initiative in programming the rules, endow the agent with explicit knowledge specified in an abstract language, and maintain the rules over time, as habits or events change.

IBM's RAISE (Reusable Agent Intelligence Software Environment) is an example of rule-based reasoning. RAISE is the inference engine of IBM's Agent Building Environment (ABE) developer's toolkit. It can perform information flow functions: finding, searching, filtering, categorizing, storing, routing, and/or selectively disseminating information items. Prototype applications for RAISE include e-commerce shopping, customer service support, and workflow on the Web and in Lotus Notes, news, and e-mail [26].

MAGSY multi-agent rule-based system is another example. The Kernel of an agent in MAGSY is a forward-chaining rule interpreter; therefore, each agent has the problem solving capacity of an expert system. In this kind of multi-agent system, the knowledge of the agents is usually structured in an object-oriented knowledge representation scheme. There is a global knowledge base, which contains the knowledge that may be accessed by all of the agents. Agents may store their identification in this global knowledge base and thus become known to all agents in the system [26].

One problem with rule-based systems is that users must keep them updated manually. These systems cannot change by themselves. A second problem is that complex sets of rules may develop conflicting rules that the agent can't resolve.

5.3 Knowledge-Based Reasoning

Knowledge-based systems (KBS) are a relatively mature aspect of artificial intelligence technology. These systems solve problems in complex application and ill-structured domains by using a large body of explicitly represented domain knowledge to search for solutions.

One can build knowledge bases based on a specific subject area or domain. These then serve as the basis for some agent inference mechanisms, including the rule-based reasoning techniques mentioned above. Usually, the developers or the knowledge engineers endow programs with information about the tasks to be performed by an agent in a specific domain, and the agent infers the proper response to a given situation. The major problem with such systems is that they require a large amount of work from the knowledge engineers. Furthermore, the knowledge of the agent is fixed and cannot be customized to the desire of individual users. In highly personalized applications, the knowledge engineer cannot possibly anticipate the best aid for each user in each situation.

In order to solve certain kinds of complex problems by knowledge-based agents, it is beneficial to create a system in which a number of Knowledge Base Agents (KBAs) cooperate and combine their problem solving capabilities. Sometimes this occurs because the problem-solving activity covers a large geographic region (such as in telecommunication networks and military applications), where different KBAs have responsibility for different geographical areas. Sometimes it occurs because different KBAs have different "specialties" to bring to the problem-solving process, similar to the co-operation among human team members. The Multiple-Agent System (MAS) paradigm has proven a popular and effective method for building a co-operating team of KBA. Each KBA in the team is constructed as a software agent, conferring abilities of autonomy, self-knowledge, and acquaintance knowledge on the KBA abilities useful for team-forming and co-operative problem solving.

5.4 Implementation

The general objective of our work is to develop an autonomous multi-agent system that retrieves, filters, integrates/conflates and validates geo-spatial data from multiple heterogeneous sources. It is assumed that the data can be stored in a number of formats, the geo-spatial databases can be relational as well as object-oriented, and different software vendors can be the sources of the database environments. Finally, the data sources are distributed and include selected proprietary databases as well as, potentially, web-based resources.

Because of the distributed and complex nature of the problem, agent technology was selected as the implementation paradigm. Intelligent agents offer several advantages over standard client-server architecture. The ability to move processing to the source of activity, i.e., a database server, reduces the network overhead. Moreover, multiple agents can simultaneously process information stored in multiple data locations. Such agents can communicate and cross-reference distributed data to support the distributed conflation process. The ability to deploy software to a remote site in an unobtrusive manner extends the functionality of the local server and allows for utilization of additional available resources. Finally, as soon as agents are deployed out of the local machine, the connection to the network can be closed (e.g., supporting the security of the system), or the local host can even be shutdown without adversely affecting the data acquisition process. A system diagram is shown in Figure 19.

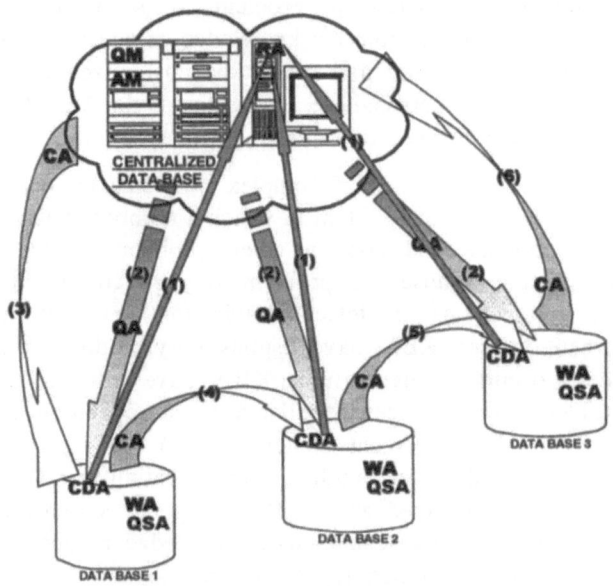

Fig. 19. Mobile agent system for distributed spatial databases.

The work presented in this chapter is implemented in the ConflationAgents (CAs). The CA is a superclass of many specialized agent classes that have extensive knowledge about their domain relevant to the conflation process. CAs are intelligent mobile agents, traveling to the feeder databases to perform conflation. The CA traverses all of the relevant databases, collecting the information and executing the knowledge-based conflation algorithm, of which fuzzy spatial relationship matching is a component. For more details on the mobile agent system interactions, please refer to [26]. As of now, a primitive implementation of the CA based on the Java Expert System Shell (JESS) has been initiated.

6 Summary and Future Work

This chapter has provided a discussion of several issues related to uncertainty in spatial relationships. We presented a brief summary of others' approaches to the topic, followed by a somewhat in-depth definition of our model based on ASGs. We showed how this model supports the definition of both fuzzy directional and fuzzy topological relationships. Several modifications of the original model were also presented, including one based on MBR refinements, one to support a specific implementation and one to handle certain cases that can produce anomalous query results. Two implementations, one for an Oracle database environment and the other based on an expert system shell were also described. We also discussed the applicability of mobile agent technology to implementation of the model for a distributed system.

Our next step is to fully implement the ASG model in a mobile agent environment. We plan to provide a prototype system that supports multiple, heterogeneous spatial databases. Fuzzy querying of spatial relationships is an essential component both for user interaction and for more complex tasks related to spatial data conflation. We believe that the ASG model or one of its variants will provide a reliable framework for a robust, working system.

Acknowledgments

We would like to thank the National Imagery and Mapping Agency, the Marine Corps Warfighting Lab, PE 0603640M, the Office of Naval Research, PE 0603238N, and the Naval Research Laboratory for sponsoring this research.

References

1. J. F. Allen. Maintaining Knowledge about Temporal Intervals. *Communications of the ACM*, 26(11):832-843, 1983.
2. J. Baldwin. Knowledge Engineering Using a Fuzzy Relational Inference Language. In *IFAC Symposium on Fuzzy Information Knowledge Representation and Decision Analysis*, pages 15-21, 1983.
3. B. Buckles and F. Petry. A Fuzzy Model for Relational Databases. *Fuzzy Sets and Systems*, 7:213-226, 1982.
4. P. Burrough and A. Frank. *Geographical Objects and Indeterminate Boundaries*, Taylor and Francis, London UK, 1996.
5. P. Burrough, P. van Gaans, and R. MacMillian. High-resolution landform classification using fuzzy k-means. *Fuzzy Sets and Systems*, 113(1):37-52, 2000.

6. M. Belgrave. The Unified Agent Architecture: A White Paper. *URL: http://www.ee.mcgill.ca/ ~belmarc/uaa_paper.html*, 1999.
7. C. Chang and T. Wu. Retrieving the Most Similar Symbolic Pictures from Pictorial Databases. *Information Processing and Management*, 28(5):581-588, 1992.
8. E. Clementini, J. Sharma, and M. J. Egenhofer. Modelling Topological and Spatial Relations: Strategies for Query Processing. *Computers and Graphics*, 18(6):815-822, 1994.
9. *CLIPS Reference Manual*, Version 6.10, August 5th, 1998.
10. M. Cobb. *An Approach for the Definition, Representation and Querying of Binary Topological and Directional Relationships between Two-Dimensional Objects*, Ph. D. Thesis, Tulane University, 1995.
11. M. Cobb and F. Petry. Modeling Spatial Data within a Fuzzy Framework. *Journal of Amer. Soc. Information Science*, 49:253-266, 1998.
12. M. Cobb, F. Petry, and V. Robinson. Special Issue: Uncertainty in Geographic Information Systems and Spatial Data. *Fuzzy Sets and Systems*, 113(1), 2000.
13. V. Cross and A. Firat. Fuzzy objects for geographical information systems. *Fuzzy Sets and Systems*, 113(1):19-36, 2000.
14. S. Dutta. Topological Constraints: A Representational Framework for Approximate Spatial and Temporal Reasoning. In *SSD'91 (Advances in Spatial Databases: 2nd Symposium)*, pages 161-180, 1991.
15. M. J. Egenhofer and R. D. Franzosa. Point-set Topological Spatial Relations. *Int. Journal of Geographical Information Systems*, 5(2):161-174, 1991.
16. Environmental Systems Research Institute (ESRI). *ARC/INFO® User's Guide: ARC /INFO® 6.0 Data Model, Concepts and Key Terms*, Environmental Systems Research Institute, Redlands, CA, 1992.
17. A. Galton. *Perturbation and Dominance in the Qualitative Representation of Continuous State-Spaces*, Technical Report 270, Department of Computer Science, University of Exeter, Exeter, 1994.
18. C. Giardina. *Fuzzy Databases and Fuzzy Relational Associative Processors*, Technical Report, Stevens Institute of Technology, Hoboken NJ, 1979.
19. M. Goodchild and S. Gopal (Eds.). *The Accuracy of Spatial Databases*, Taylor and Francis, Basingstoke, UK, 1990.
20. H. W. Guesgen. *Spatial Reasoning Based on Allen's Temporal Logic*, Technical Report TR-89-049, International Computer Science Institute, Berkeley, CA, 1989.
21. H. P. Kriegel, M. Schiwietz, R. Schneider, and B. Seeger. Performance comparison of Point and Spatial Access Methods. In: A. Buchmann, O. Gunther, T. Smith, and Y. Wang (Eds.), *Design and Implementation of Large Spatial Databases*, LNCS 409, pages 89-114, Santa Barbara, CA, Springer-Verlag, 1989.
22. D. MaGuire. An Overview and Definition of GIS. *Geographical Information Systems: Principles and Applications*, D. MaGuire, M. Goodchild, and D. Rhind (Eds.), 1:9-20, Longman, Essex GB, 1991.
23. A. Mukerjee and G. Joe. A Qualitative Model for Space. In *AAAI-90 (8th National Conference on Artificial Intelligence)*, pages 721-727, 1990.
24. M. Nabil, J. Shepherd, and A. H. H. Ngu. 2D Projection Interval Relationships: A Symbolic Representation of Spatial Relationships. In *SSD '95 (Advances in Spatial Databases: 42nd Symposium)*, pages 292-309, 1995.
25. D. Papadias and Y. Theodoridis. Spatial Relations, Minimum Bounding Rrectangles, and Spatial Data Dtructures. *Int. J. Geographical Information Science*, 11:111-138, 1997.
26. S. Rahimi, A. Ali, and D. Ali. An Investigation on Intelligent Software-Agent Technology. In *IEMS and IC&IE Joined Int. Conference*, Cocoa Beach, Florida, 2001.
27. V. Robinson. Implications of Fuzzy Set Theory for Geographic Databases. *Computers, Environment, and Urban Svstems*, 12:89-98, 1988.

28. V. Robinson. Interactive Machine Acquisition of a Fuzzy Spatial Relation. *Computers and Geosciences*, 6:857-872, 1990.

29. V. Robinson and A. Frank. About Different Kinds of Uncertainty in Geographic Information Systems. In *AUTOCARTO 7*, 1985.

30. H. Samet. Applications of Spatial Data Structures: Computer Graphics, Image Processing, and GIS. Reading, MA: Addison-Wesley, 1989.

31. J. Sharma and D. M. Flewelling. Inferences from Combined Knowledge about Topology and Direction. In *SSD'95 (Advances in Spatial Databases: 42nd Symposium)*, pages 279-291, 1995.

32. D. Stoms. Reasoning with Uncertainty in Intelligent Geographic Information Systems. In *GIS'87 (2nd Annual Int. Conf on Geographic Information Systems)*, pages 693-699, American Soc. for Photogrammetry and Remote Sensing, Falls Church VA, 1987.

33. F. Wang. A fuzzy grammar and possibility theory-based natural language user interface for spatial queries. *Fuzzy Sets and Systems*, 113(1):147-159, 2000.

34. S. Winter. Topological Relations between Discrete Regions. In *SSD'95 (Advances in Spatial Databases: 42nd Symposium)*, pages 310-327, Portland, ME, USA, August 1995.

35. S. Winter. Uncertainty of Topological Relations in GIS. In *ISPRS Commission III Symposium*, pages 924-930, 1994.

36. R. M. Wygant. CLIPS – A Powerful Development and Delivery Expert System Tool. *Computers in Engineering*, 17(1/4):546-549, 1989.

37. L. Zadeh. Test-Score Semantics for Natural Languages and Meaning Representation via PRUF. *Empirical Semantics*, B. Rieger (Ed.), pages 281-349, Brockmeyer, Bochum, GR, 1981.

Using Fuzzy Spatial Relations to Control Movement Behavior of Mobile Objects in Spatially Explicit Ecological Models

Vincent B. Robinson

Department of Geography
University of Toronto
Mississauga, ON L5L 1C6 Canada
vbr@geomant.erin.utoronto.ca

Abstract. Spatial relations are fundamental to the modeling of spatially explicit ecological processes. An information-based framework relates geographical information system (GIS) data to object representations of individual mobile animals. It is used to organize a discourse on the incorporation of fuzzy logic into spatially explicit, individual-based ecological models of animal movement across a landscape of habitat. Spatial relations such as proximity, direction, overlap, and containment are used in a spatial reasoning process to control movement of individual animal objects over a landscape. It is shown that an animal's perceptual range can be specified as a function of proximity to an animal object and used as a fuzzy spatial constraint region for object queries operating over a landscape database. The role of fuzzy relations in models of habitat evaluation is addressed. The potential use of fuzzy spatial relations in modeling movement behavior primarily associated with foraging is demonstrated. It is shown that spatially explicit ecological modeling is a complex domain rich in the potential for intelligent applications using fuzzy spatial relations

Keywords. Geographic information systems, fuzzy spatial relations, mobile objects, ecological modeling.

1   Introduction

This paper illustrates how fundamental spatial relations like distance, direction, and topology can be represented as fuzzy relations that are meaningful in the context of spatially explicit ecological models. There is special emphasis on using fuzzy spatial relations to control the behavior of animal objects during the simulation run of a spatially explicit ecological model. First, a general framework is discussed that illustrates that these models can be quite complex and are a microcosm of a much broader range of issues regarding the incorporation of fuzzy sets in spatially-explicit models. This paper first briefly considers the issues raised when using fuzzy logic in the creation of the spatially explicit landscape database.

The role of fuzzy spatial relations in the habitat evaluation process is illustrated with a simple example. An illustrative example of one approach to modeling the fuzzy control of foraging movements over a landscape of habitat highlights the role that fuzzy spatial relations play in controlling the behavior of objects in such spatially explicit models. The concluding section discusses, in more general terms, further potential research themes in fuzzy sets, particularly fuzzy spatial relations, that can contribute to development of spatially explicit ecological models that incorporate the specification and manipulation of fuzzy spatial relations.

1.1 Information-Based Approaches to Ecological Modeling

Ecologists have begun to use spatially explicit models to examine connections between landscape patterns and species viability [21]. These efforts often link behavior ecology and landscape-level ecological processes in an information-based approach to modeling the movement/dispersal of animals [13]. In fact, it has been argued that the difficulties of incorporating different levels of habitat hetero-

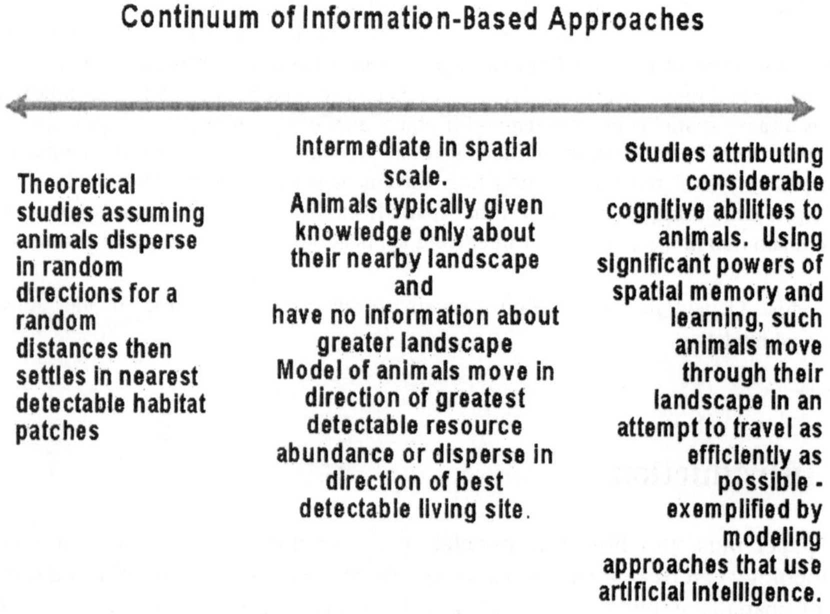

Continuum of Information-Based Approaches

Theoretical studies assuming animals disperse in random directions for a random distances then settles in nearest detectable habitat patches

Intermediate in spatial scale.
Animals typically given knowledge only about their nearby landscape and
have no information about greater landscape
Model of animals move in direction of greatest detectable resource abundance or disperse in direction of best detectable living site.

Studies attributing considerable cognitive abilities to animals. Using significant powers of spatial memory and learning, such animals move through their landscape in an attempt to travel as efficiently as possible - exemplified by modeling approaches that use artificial intelligence.

Fig. 1. Continuum of information-based approaches
to modeling movement/dispersal of animals (based on [13]).

geneity, individual differences and local interactions into mathematical models suggest that a general theory of individual-based ecology is impossible and that we have no choice but to develop detailed computer-supported models [14]. Using Figure 1 as a referent, this paper will address issues related to developing models that tend to fall somewhere between the intermediate level and those studies attributing considerable cognitive abilities. In particular, the models considered here generally assume the animals have knowledge about their nearby landscape and no information about greater landscape, but often include a degree of spatial memory and use methods/techniques drawn from the broad field of artificial intelligence. At this region of the continuum, information about the landscape is essential to the model-building enterprise. Thus, it is not surprising that the use of models that connect animal populations to "maps" stored in geographic information system are now a prominent feature in the field of conservation biology [21]. Generally speaking, such spatially-explicit population models use a GIS database to configure the layout of available habitat and then apply a detailed simulation of individual organisms moving through the landscape (see Figure 2). Due to enabling spatial and computational technologies these models allow one to describe a landscape in as much detail as a GIS database can support [e.g., 11] and because they are individual-based models (IBMs), they can represent realistic behavior with parameters that reflect mechanisms thought to be responsible for a species being at risk in fragmented habitat [21].

There remain uncertainties regarding foraging, exploring, and conspecific interaction that will need to be confronted through development of new models [23] and as a result of field observations. However, it was not the purpose of this work to present a comprehensive model but rather to illustrate how and where fuzzy sets, especially fuzzy spatial relations, might be useful in constructing spatially explicit ecological models.

1.2   Framework for Spatially Explicit Ecological Modeling

Figure 2 presents in a generalized fashion the major components of a spatially explicit model and the relationship between each of them. Of critical importance in all the models is some representation of the landscape. In foraging models these will generally be representations of how habitat quality, especially food density/quality, varies over space. It may be derived from extensive field observations or inferred from sources such as land cover maps, or even satellite remote sensing data. Here this landscape will be treated as a spatial database from which the animal objects will receive information about their surroundings. Generally speaking, the landscape is not modeled as changing in any significant manner during the simulation process. This is usually due to the already complex nature of the model that introducing this level of complexity would obscure behaviors that are the focus of the research effort. Animal objects pose spatial queries to the landscape to acquire information. That information is then processed to determine the specifics of which movement behavior to pursue. Later two kinds of movements, controlled using fuzzy logic, will be illustrated - foraging and exploratory. Both from a simple reasoning process that leads to an animal object's spatially explicit decision as to where to move to next.

Results of one simulation study suggested that errors in dispersal parameters have much larger consequences for predicting dispersal success than did errors in landscape classification [21]. However, it was subsequently noted that in spatially explicit population models, landscape classification can affect demographic processes through an effect on patch carrying capacity (i.e., habitat quality). This could increase the importance of landscape classification in such models [23].

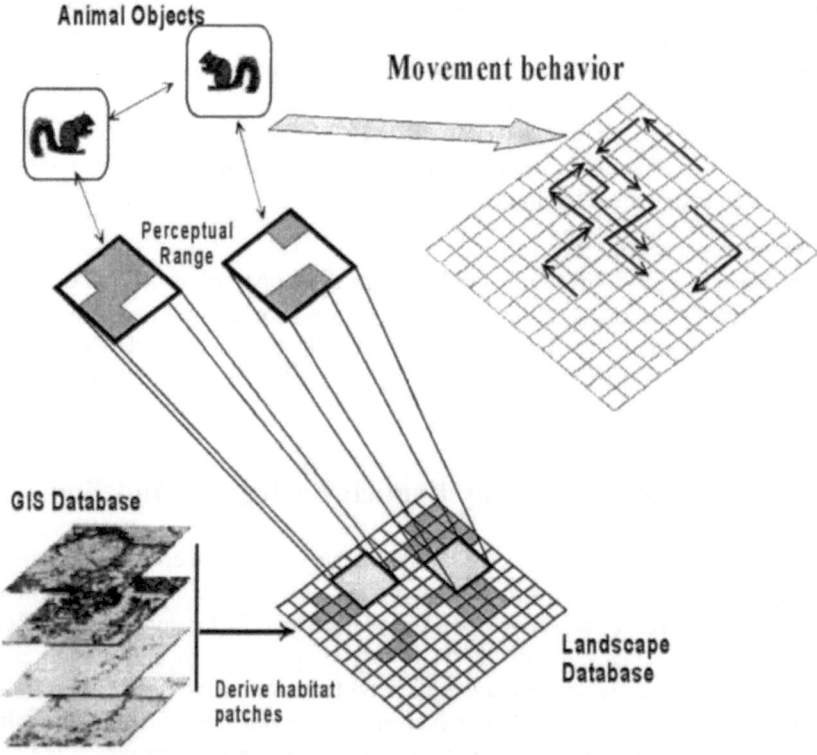

Fig. 2. Conceptual framework showing relationship between geographic information system (GIS) database, the landscape database describing the spatial distribution of habitat, animal objects, their perceptual range, and their movements over the landscape.

Keeping these implications in mind, this paper first briefly considers the issues raised when using fuzzy logic in the creation of the spatially explicit landscape database. The use of fuzzy spatial relations will be highlighted. Following consideration of those issues the focus will be on illustrating the potential use of fuzzy spatial relations in modeling movement behavior primarily associated with foraging.

2 Modeling Habitat Landscape

Fuzzy set theory has been used to provide a definition of the niche concept in an attempt to address, among other issues, the ambiguity characterizing the boundaries of niche space [5]. Although Cao [5] presented fuzzy definitions of niche width and distance, a formulation linking such work to spatially explicit habitat models is still lacking. This is not surprising considering the difficulty in determining niche space much less how it may vary over space. One approach is to use a habitat suitability index (HSI). The reliability of habitat units is therefore dependent on a well-defined and accurate HSI. However, in most cases HSI models are constructed from expert opinion which is notoriously variable. In addition, the opinions may not be comparable [4]. This led to a study on the use of fuzzy numbers to solve for the distribution of uncertainty around an HSI. Such habitat models can be used in conjunction with GIS to develop a representation of the habitat landscape.

2.1 Fuzzy Spatial Relations in Habitat Evaluation

Spatial relations are sometimes incorporated in the models at the habitat evaluation level. Table 1 summarizes a few such relations from selected recent studies using fuzzy sets in the modeling of habitat quality. Note that in the studies in Table 1 that the spatial relations are primarily proximity relations. Depending on how it is derived, the density measurements are a function of topological and/or proximity relations. The specification of the membership functions

Table 1. Selected examples of elements of habitat models that are a function of spatial relations. These studies all used fuzzy sets.

species: Florida Scrub-Jay [4]

    distance from scrub oak ridge

    distance from ruderal area

species: foliage-dwelling spiders [12]

    margin width

    margin density

species: deer [17]

    distance to water

    road density

associated with each of these spatial relations is usually based on expert opinion and existing literature. Other studies [e.g., 3] do not incorporate spatially explicit relations in their fuzzy habitat model. However, when spatial relations are used in a fuzzy model of habitat they can play an important role, but one that is secondary to that of forage and cover components.

In Figure 2 the landscape upon which animal objects operate is a representation of how habitat varies over space. Basic data from a GIS database is typically used to determine some kind of habitat representation scheme. Often the GIS database is a set of maps that have been compiled from field measurements, aerial photos, or analog maps. It is common at the landscape scale that the habitat 'map' is derived from land cover classifications of remotely sensed data like that from the Landsat series of satellites. In either case, rules are developed that evaluate the GIS data in terms of habitat requirements for a particular species.

Each of these approaches is briefly discussed below. The use of expert opinion and literature to construct a fuzzy habitat model is illustrated by considering an object-oriented model of deer evaluation habitat [17]. This is followed by a discussion of the use of remote sensing data as the basis for deriving habitat models.

2.2 An Example of Fuzzy Habitat Evaluation

Using an object-oriented approach to model how deer evaluate their habitat Rickel et al. [17] initially constructed a GIS coverage that consisted of polygons that were coded as type of vegetation derived from timber stand and range analysis maps. Each of these polygons were linked to nine habitat parameters (Table 2). Using a definition of 'good habitat' fuzzy sets for each of the habitat parameters were developed by developing a mapping to fuzzy memberships using rules for defining fuzzy sets such as good forage, good cover, and cool season growers.

Of particular interest are the spatial relations used in the habitat model. The most explicit is the spatial relation of ***near_water***. It is specified by a declining S-curve (Figure 3). The specific parameters of the S-curve were determined using expert opinion and the literature. For the general deer object the curve represents the proposition that the most effective distance to water for deer habitat is within one mile while locations three miles or more are least

Table 2. Habitat parameters used in a model of deer habitat evaluation [17].

vegetation
range condition class ratings
browse composition ratings
browse availability ratings
browse density rating
tree size
canopy closure
distance to water
road density

effective. Another spatially-related parameter was road_density. Road_density was the percent area of a polygon that lies within 1/8 mile of a road. Thus, it is a combination of both proximity to a road relation and the overlap of the buffer around each road and habitat polygons.

The use of the compensatory mean to arrive at a fuzzy membership in the set of **Natural_habitat_quality** illustrates the ranking given to the spatial relation of **near_water** [17]. The compensatory mean (see Equation (1)) shows that their model assumes that forage and cover are the two most important factors followed by the spatial relation **near_water**.

$$\mu_{Natural\_habitat\_quality}=(((((\mu_{forage}+\mu_{cover})/2)+\mu_{near\_water})/2)+\mu_{cool\_season\_growers})/2) \qquad (1)$$

For areas that have been disturbed by the construction and use of roads the **Natural_habitat_quality** was used in conjunction with the **road_density** in (2).

$$\mu_{Disturbed\_habitat\_quality} = \mu_{Natural\_habitat\_quality} \bullet \mu_{road\_density} \qquad (2)$$

In this model the doe object inherited all the functions from the general deer object with the **near_water**, plenty of cool season growers functions modified to reflect preferences of the does. Similarly, the fawn object inherits doe functions with cover and **near_water** modified. The spatial relation of **near_water** is the function that varies across all three deer object types.

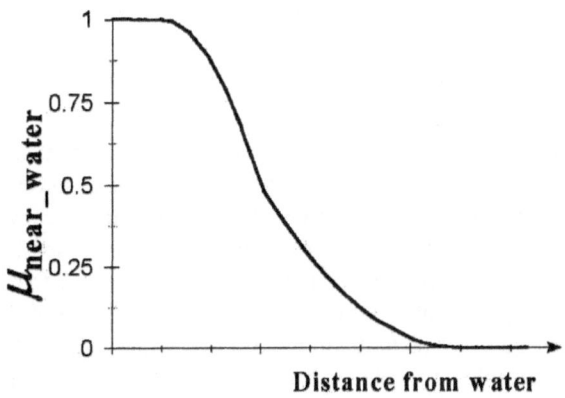

Fig. 3. Idealized S-curve for **near_water** set
for general deer object (adapted from [17]).

2.3   Land Cover Classification and Habitat Modeling

At the landscape level the use of land cover 'maps' derived from satellite remote sensing has become common place. For example, an individual-based dispersal simulation model was developed to investigate the behavior of dispersing individuals and measure its influence on immigration/emigration rates between habitat islands in an heterogeneous landscape [10]. Realistic landscapes were used. The realistic landscapes were land cover classifications based on satellite remote sensing data. Each grid cell was classified as belonging to one of a possible seven land cover classes with the forest class being the 'preferred' habitat. However, it is well-known that a pixel in a remote sensing dataset may encompass more than one 'pure' land cover class. Since Robinson and Thongs [20] showed how fuzzy sets could be applied to such a mixed pixel problem, there has been considerable interest in using fuzzy sets in the land cover classification process [e.g., 9,25] Although the end result may be a land cover map in which a pixel is ultimately assigned to only one particular class, there is a stage in the process where each pixel is assigned a fuzzy membership value. Typically some kind of defuzzification rule is used to allocate a pixel to a single class. Alternatively, such fuzzy classification information may used to develop representations of fuzzy objects. The Fuzzy-Crisp object (FC-object) described by Cheng et al. [6] has a crisp internal core but a fuzzy spatial extent. The FC-object is quite similar to the idea of habitat patches having edges (fuzzy spatial extent) and an interior (crisp core). However, in this case the Fuzzy-Fuzzy object (FF-object) which has a fuzzy spatial extent and a fuzzy thematic interior may be more appropriate to represent the level of information provided by fuzzy land cover classification using remotely sensed satellite data.

Considerable attention has been given to improving the classification accuracy of land cover maps derived from remotely sensed data. Some of those approaches have made use of fuzzy set theory, particularly Bezdek's fuzzy c-means algorithm [2]. However, the goal frequently remains that of allocating a pixel to a single land cover class. This objective is often driven by the demands of commercial GIS software that do not easily support the representation of fuzzy spatial objects in the GIS database. When a location, or spatial object, is allocated to a single land cover class we lose a significant amount of information that could otherwise be used by knowledge based systems that can handle fuzzy knowledge. For example, Figure 4 illustrates a simple example of how output from a fuzzy classification process is defuzzified. Usually a max operator is used to arrive at a spatial dataset in which each cell is coded as a member of a single land cover class. Note that there are several instances in Figure 4 where rules for deciding 'ties' will have to be used to determine which land cover class it is allocated to. Often this is done by evaluating the neighborhood of the pixel in question. Although this may arrive at an acceptable land cover dataset for subsequent use, we have lost information about those cells that had a high membership in both forest and grassy cover types. This could be important for species that prefer forest/grassy edge habitat. This implies that representing land cover objects using fuzzy set concepts that can subsequently be used by animal objects may lead to more meaningful models.

Several studies have illustrated the utility of fuzzy rule-based models in explaining the population dynamics of particular species [e.g., 3,12]. This suggests that an integration of fuzzy object representation in a GIS database can now be integrated with a knowledge based GIS. Examples of representing fuzzy attributes and relations of objects in a knowledge based GIS illustrate how fuzzy extensions of an object definition language could represent the fuzziness inherent in many kinds of geographic data. On the other hand, those examples also suggest that the introduction of fuzzy membership grades may affect the inheritance of attribute values, constraints, and methods. Therefore, it is not clear what the general rules governing the inheritance should be. It has been suggested that at this stage in our understanding of fuzzy object management the rules governing fuzzy inheritance may be context specific. In other words, the inheritance rules governing inheritance of attribute values, constraints, or methods may be depend on the specific semantics of the problem domain [8,19].

For the purposes of these spatially explicit ecological models it may be sufficient to represent a landscape populated by fuzzy land cover objects that have

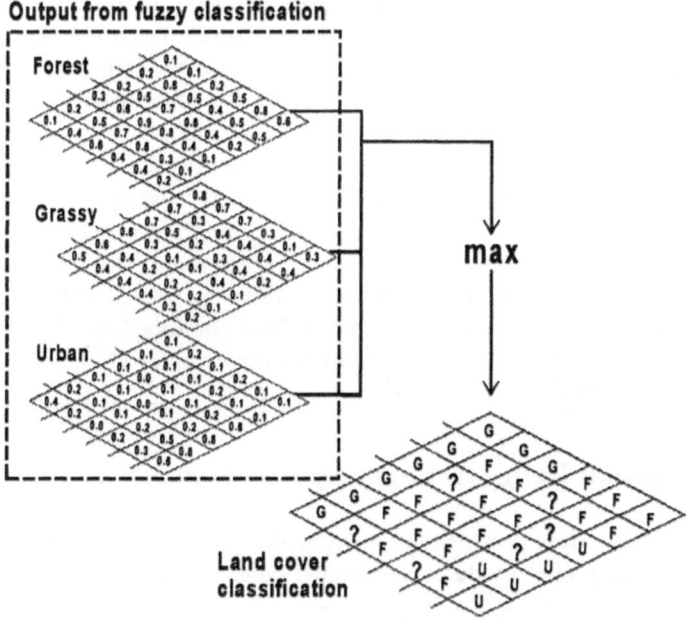

Fig. 4. Example using the max operator for defuzzification
of results from a fuzzy land cover classification process.

knowledge of their relationship to one another. However, this approach would necessitate a further level of processing whereby the animal objects would query the landscape for basic information and then process that information to arrive at an assessment of 'habitat'. Such a query would need to make use of spatial relations.

3   Fuzzy Control of Spatial Movement

Recall that results of one simulation study suggested that errors in dispersal parameters have much larger consequences for predicting dispersal success than did errors in landscape classification [21]. The previous section discussed land cover classification and habitat modeling in some depth because it is central to an animal object's decision whether or not to disperse to, through or forage in a particular location. However, the individual objects have to make spatially explicit decisions about when to move, where to move. The where decision is not only distance-based but may also be direction-based. Here the use of fuzzy sets, especially fuzzy spatial relations, in controlling dispersal or forage behavior over space is discussed.

A computing environment that supports development of spatially explicit individual-based modeling should support, among other requirements, mobility, evaluating surrounding individuals, interactions with other individuals, acquiring and maintaining knowledge [24]. Figure 5 illustrates one approach to spatially explicit simulation of animals where the AnimalInfo class for accessing/changing facts about a given animal type provides a common area for animals and animal support classes to communicate with each other. It also provides means for accessing facts associated with all other animal entities. LandInfo provides the animal's view of its environment by retrieving information from the GIS. In many simulations, an animal is able to collect information, or view, its surroundings within a certain perceptual range. It is, in theory, possible to combine both the AnimalInfo and LandInfo object classes so that one object class handles all queries for information from individuals.

3.1   Perceptual Range as Fuzzy Spatial Relation

An important controlling parameter in many spatially explicit simulation models is the perceptual range of individuals. Perceptual range is the distance from which a particular landscape element can be perceived. An animal's perceptual range represents its informational window onto the larger landscape. The spatial limits of the perceptual range determine how much of the area surrounding the individual can be perceived [28] in terms of habitat quality and other conspecifics.

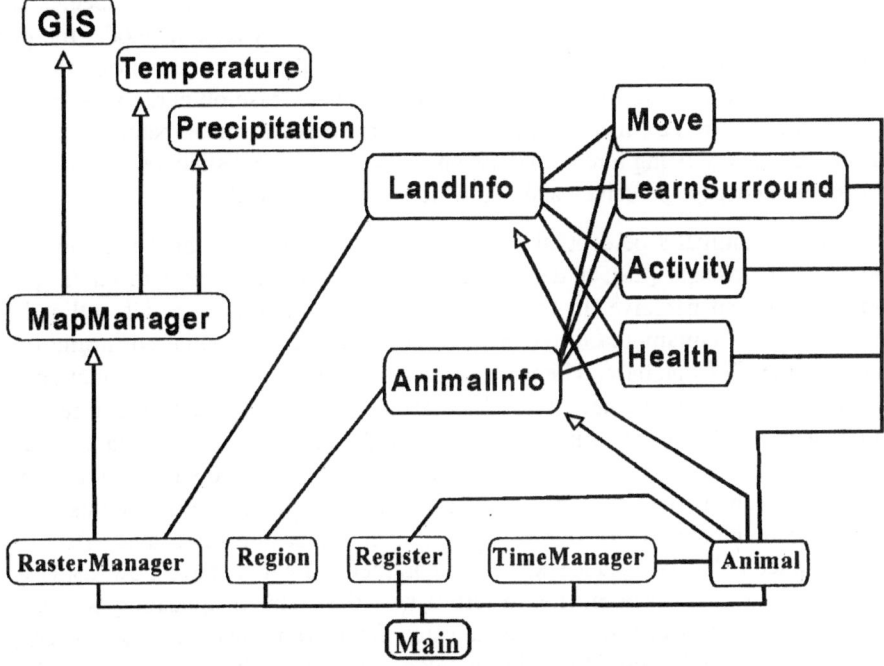

Fig. 5. An example of an objet class hierarchy used to model movement behavior of individual animals over a landscape (adapted from [24]).

Models incorporating perceptual range have typically specified them as crisp sets [1,16,23,24,28]. However, like the example of margin widths used by Kamplicher et al. [12] in their study of species density of foliage dwelling spiders using a fuzzy rule-based model, applying crisp boundaries to the concept of a perceptual range may not be biologically meaningful. Consider the research on the landscape level perceptual abilities of forest scriurids where results suggest that the perceptual range of three sciurids varied as summarized in Table 3 [27].

Table 3. Estimates of perceptual range for three sciurids based on results from [27].

Species	Perceptual Range
Eastern chipmunk (*Tamias striatus*)	between 120 and 180 meters
Gray squirrel (*Sciurus carolinensis*)	between 300 and 400 meters
Eastern fox squirrel (*Sciurus niger*)	between 300 and 500 meters

It is important to note the variation among different, yet similar, species and the width of the edge of perceptual range. If in a spatially explicit model the perceptual range of our chipmunk objects were set to 120 meters the biological interpretation would be that the chipmunk could not in any fashion perceive habitat at 120.01 meters, or 121 meters. Thus, using a crisp set form of perceptual range means that we are being forced to draw an artificially sharp distinction between that portion of landscape perceived and that which is not perceived.

Since an animal's perceptual range represents its informational window onto the larger landscape [28], it determines how much of the area surrounding the individual it can perceive. In the spatially explicit simulation model outlined in Figure 2 this is tantamount to the perceptual range being a spatial constraint on a query to the GIS database. There are a wide variety of membership functions to choose from. Each will characterize the perceptual range as a fuzzy set. It remains to be seen which function may be the most suitable given behavior data about a particular species. For example, if we let $\mu_P(x)$ be the degree to which a location is a member of the perceptual range, the perceptual range could be defined using (3) where the membership function is defined over a measure of distance from the animal object, that is distance x, where parameter θ will control how $\mu_P(x)$ approaches the crossover point (i.e., where $\mu_P(x) = 0.5$). This is appealing because it is more straightforward to arrive at a value for β from field data, and perhaps expert experience. In addition, field data can also suggest how large θ should be. Using this membership function each location, or spatial object, in the landscape could be coded with a membership value indicating the degree to which that location is in the perceptual range of that individual at a particular time step.

$$\mu_P(x) \;=\; \begin{cases} 1/[1+((x-\beta)/\theta)]^2 & \text{if } x \ge \beta \\ 1 & \text{if } x < \beta \end{cases} \tag{3}$$

On the other hand, a trapezoidal function may be easier for some to parameterize from field observations, or experience. For example, in (4), θ is a parameter determining the rate at which the membership decreases with an increasing distance from β (see Figure 6).

$$\mu_P(x) \;=\; \begin{cases} 1 & \text{if} & x \le \beta \\ \theta(\beta-x)+1 & \text{if} & \beta < x < \beta+1/\theta \\ 0 & \text{if} & \beta+1/\theta \le x \end{cases} \tag{4}$$

Since animal objects must query the GIS database to obtain their information about their landscape we may treat the role of perceptual range as a significant constraint in a spatial query the animal object poses to a landscape database. In particular the fuzzy set P can be used to parameterize a range query on a spatial database. As a first approximation, one can use an α-cut such that:

$$^{\alpha}P \;=\; \{x \mid \mu_P(x) \ge \alpha\}. \tag{5}$$

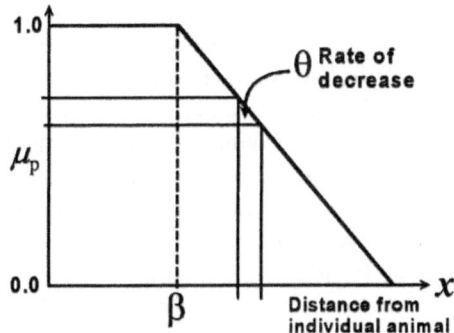

Fig. 6. Example of trapezoidal function used to specify membership in the perceptual range set (μ_p).

The information search strategy can be fine tuned using decreasing α-cut to increase the size of search window (see Figure 7). The criteria for relaxation of α would be a function of whether or not any acceptable destination was perceived. For example, if as a result of a search of the area in belonging to $^{1.0}P$ there was no suitable destination found, then $^{1.0}P$ would be relaxed to $^{0.5}P$. If any suitable destinations were perceived within the region of $^{0.5}P$ the relaxation of α would cease and the animal object would evaluate the potential destinations based on criteria discussed below, then make its move.

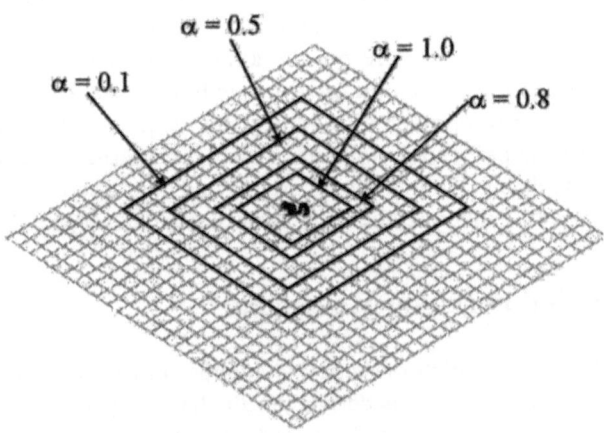

Fig. 7. Example of how changing α-cut affects the spatial extent of an individual's perceptual range.

It is easy to see that unless we are studying animals with very large perceptual ranges, this would be computationally inefficient, especially when the perceptual range is very small in relation to the landscape database being queried. Keep in mind that this is occurring in the context of many animal objects operating over the same landscape during the same time period. Thus, efficient spatial retrieval can become a significant issue when working with large complex landscapes. The minimum bounding rectangle (MBR) is a common approach when addressing retrieval issues through the logic of the spatial model. If MBRs are used but are maintained as the crisp approximations there may be difficulty in mapping from fuzzy spatial queries to crisp MBR approximations. An approach that uses distance constraints based on fuzzy spatial similarity relations which use distance from the centroid of spatial objects [e.g., 14] may be a bit crude for the purposes of such a model because it is sufficient for the animal to perceive just the edge of the landscape object. In other words, the centroid of the object maybe some distance from its edge whereas the edge is what is important. It is easy to see how this approach can be adapted to handle the increased granularity of a database that supports such queries. Another approach set in a fuzzy framework uses abstract spatial graphs (ASG) to represent a transformation of 2-dimensional space into 0-dimensional space. The ASG approach employs MBRs to represent features. It addresses some of the challenges posed in the modeling of topological spatial relations using MBR representations by not only representing topological and directional relationships, but also supplemental information needed for fuzzy query processing. For example, each node in an ASG has weights that can be used to define fuzzy qualifiers for a query. Thus, an approach such as ASG could be used to extend a crisp spatial retrieval model to one that explicitly incorporates fuzziness in the spatial graph [7]. Another recent approach uses configuration similarity to address retrieval problems associated with a query like - 'find all configurations where an object x is about 5km northeast of y which in turn is inside z' [15]. Although these are not the kinds of spatial queries made by the animal objects to the landscape, it does illustrate how spatial database query problems may need to be reformulated to address the demands of these spatially explicit models. This is particularly true when there are a large number of individual objects asking the database about the landscape and their queries all have a spatial constraint representing the fuzzy limits of their perceptual range.

3.2 Controlling Foraging Movement

The GIS database over which the query is applied may either consist of the raw data from which habitat quality is derived, or the habitat quality may already have been determined across the whole of the landscape. In the latter case, it is simply a case of retrieving the membership values for each object. If H is the fuzzy set "good habitat" then let $\mu_H^i(x)$ be the membership grade of object i at distance x. For those objects located in $^{1.0}P$, it is assumed that the preferred destination is the closest object with the maximum membership grade. However, a significant question is how low should the maximum $\mu_H^i(x)$ be before relaxing the α-cut to search a greater area. A reasonable choice would seem to be $^{0.5}H$. So, if at any point in the information retrieval process Σ Count $(^{0.5}H \geq 1)$ then the relaxation ceases and the animal object evaluates the information retrieved. As one approaches the edge of the perceptual range the distance may have an effect on an animal's evaluation of habitat. In other words given objects with the same grades of membership but the one near the edge of the perceptual range will be evaluated with an overall lower final value. This is admittedly a crude way of incorporating possible effects of distance on an animal's perceptual ability, but it will suffice for now. Using the algebraic product as our intersection function, the degree to which an object belongs to the fuzzy set of destinations is defined as:

$$\Gamma = H\,I\,P = (\mu_H^i(x) \cdot \mu_P(x)) \tag{6}$$

where μ_Γ^i is the grade of membership of spatial object i in fuzzy set Γ. After the animal object has evaluated the information within its perceptual range, either it moves to the landscape object that has the highest μ_Γ^i, or, if there is more than one potential location, then it moves to the closest one. Note that this rule may be refined with further research to determine if tradeoffs between habitat quality versus nearness would have any significant effect on patterns of behavior.

3.3 Controlling Exploratory Movement

What happens if the process concludes with $\Sigma Count(^{0.5}H < 1)$, which means no "acceptable" habitat was found? The definition of "acceptable" could be changed to include those objects with grades less than 0.5. Perhaps a more realistic approach would be for the animal object to switch from foraging/dispersal behavior to exploratory behavior. Since the animal is primarily gathering spatial information about the landscape during its exploratory behavior, it may use similar rules for movement but will relax constraints driven by the presence of forage.

Although a random search strategy could easily be implemented, it has been noted that mammals tend not to forage as a random process [23] as was the case for hummingbirds. A variation on the directional search strategy [1] modeled in a study of hummingbird foraging behavior could be used to develop a fuzzy logic for an animal object's decision on which destination to move to during its exploratory behavior. The basic idea of this strategy is that the animal object will continue on a straight patch until it encounters acceptable habitat. Here is another aspect of an animal object's behavior that may be inherently fuzzy. For example, consider the case where an animal object arrives at its present location by traveling in a direction of 310° which is generally southeast but to some extent south as well. Even though the directional search strategy seems straightforward, it may be implausible to think an animal object will display such precision in its direction movement. That is to say, why not move 309° instead of 310°? One straightforward approach is to specify a fuzzy membership function for ***direction to move*** which is based on the previous movement as in Figure 8. Let μ_Φ^i be the grade of membership of spatial object i in the fuzzy set ***direction to move*** (Φ).

When we modeled foraging movement behavior the landscape was evaluated for its membership in 'good habitat', the same basic landscape data could be evaluated for its membership in a set that can best be described as 'suitable destination.' For example, a sciurid like that studied by South [23] would consider an object without pine cones as being not habitat, but may consider a spatial object with shrubby vegetation to be a suitable destination during its movement because it offers some measure of cover while it evaluates the landscape for its next movement. For the sake of this example let us say that Ψ is the fuzzy set of suitable destinations with μ_Ψ^i being the membership grade of spatial object i in Ψ.

For each spatial object i within in the perceptual range a membership in the 'destination' could be defined as:

$$\Lambda \;=\; \Phi\, I\; \Psi \;=\; (\mu_\Phi^i \cdot \mu_\Psi^i) \tag{7}$$

and among those spatial objects with the maximum μ_Λ the farthest from the current location will be the destination.

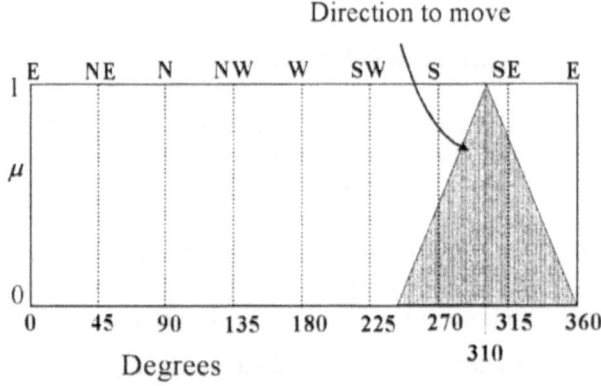

Fig. 8. Example of the fuzzy set *direction to move*. The value of 310 is based on the last direction the animal object moved to arrive at its present location.

3.4 Spatially Explicit Conspecific Interactions

Some species are strongly territorial while others are less so. Some simulations using crisp rules assume that if a conspecific is already located in a potential destination then that destination is avoided [e.g., 23]. This presumes perfect knowledge on the part of the animal object of the location of all conspecifics within its perceptual range. Since the perceptual range of an animal object has been fuzzified, it is possible to construct functions that may account for various uncertainties regarding the avoidance of conspecifics.

Research on spatial queries that formalize the concept of *overlap* could be useful in determining which animal object's perceptual range overlaps with the reference animal object. Rather than use the exact point location of an animal object their perceptual ranges can be used as a means of generally locating them. Let $^{0.8}P_i$ be the perceptual range as defined above for animal object i and $^{0.5}P_r$ be the perceptual range of the reference animal object, r. Using the formal methods to determine *overlap* in spatial queries [7,15] an efficient method can be found to arrive at:

$$T = \ ^{0.5}P_r \ I \ \left\{ Y_i \ ^{0.8}P_i \right\} \tag{8}$$

which should be all the locations (or spatial objects) where the perceptual range of the reference animal and surrounding conspecifics intersect, using the 0.5-cut. Depending on whether this is foraging or exploratory behavior, Λ or Γ is used directly if $\Sigma Count(\ T\ )=0$. Otherwise the proximity of conspecifics is taken into account when evaluating destinations. Using the perceptual range memberships a fuzzy set Ξ that describes the degree to which a location (or spatial object) is *near_conspecific*. It could be constructed such that within those locations covered by T the membership grade would be:

$$\mu_\Xi \ = \ \mu_P^r \bullet \ \left\{ Y_i \ \mu_P^i \right\} \tag{9}$$

Reformulating for the case of exploratory behavior we can say that for each spatial object i within in the perceptual range a membership in the 'destination' could be defined as:

$$\Lambda' \ = \ (\neg\Xi)\,I\,[\Phi\,I\,\Psi] \ = (1 - \mu_\Xi^i) \ \ (\mu_\Phi^i \cdot \mu_\Psi^i) \tag{10}$$

and among those spatial objects with the maximum $\mu_{\Lambda'}$ that spatial object farthest from the current location will be the destination. Similarly for the case of foraging behavior:

$$\Gamma' \ = \ (\neg\Xi)\,I\,[H\,I\,P] \ = (1 - \mu_\Xi^i) \ \ (\mu_H^i \cdot \mu_P) \tag{11}$$

Once the animal object has moved to a new location it will begin the process of evaluating the landscape within its perceptual range. If suitable forage is perceived by the animal object, then it will engage in foraging behavior, exploratory search behavior.

Keep in mind that both Λ and Γ are used in the general context of foraging behavior. Whereas animal objects may seek to avoid conspecifics when foraging due to competition for food resources, there may be a conspecific attraction factor at play when modeling dispersal behavior. This suggests in the case of dispersal:

$$\Lambda^{dispersal} \quad = \quad \Xi \, I \, [\Phi \, I \, \Psi] \quad = \mu_{\Xi}^{i} \quad (\mu_{\Phi}^{i} \cdot \mu_{\Psi}^{i}) \tag{12}$$

could be used to evaluate those locations with the highest $\mu_{\Lambda^{dispersal}}$ and farthest from the its current location. This also highlights the complexity of developing a model that takes both foraging and/or dispersal behavior into account.

4   Discussion

This work has shown how fundamental spatial relations such as near, direction, overlap and/or containment can be represented as fuzzy relations in the context of spatially explicit ecological models. In particular it was shown how fuzzy spatial relations can be used to model the controlling reasoning process of mobile animal objects moving over a landscape. Since the perceptual range governs the flow of information to an animal object, it is of particular importance in the reasoning process of an individual animal object. The information-based framework allows us to model the perceptual range as a function of fuzzy spatial relation and use it as spatial constraint when the object queries the landscape database for information about its surroundings. Thus, fuzzy spatial relations are used in combination with attribute information to formulate a decision about which spatial object to move to as part of a foraging or exploration movement.

This exploration of spatially explicit models has revealed a rich domain where fuzzy sets, especially fuzzy spatial relations, may be effectively applied. It has also been a first stage in an exploration that suggests several areas that may be particularly productive to pursue.

4.1   Fuzzy Rule-Base Models

The examples used to show how fuzzy sets can be used in the control of individual object behavior such as in foraging and exploratory movement take a straightforward approach of manipulating simply membership functions rather than developing an extensive rule-base of linguistic variables. Thus, an alternate approach that might be useful to investigate would be an explicit rule-based approach that incorporates spatial relations in both antecedent and consequent. Such an approach has been used to yield superior results in a study population dynamics of forest mice [3] and of species density in foliage-dwelling spiders [12]. Since such a rule base must cover the entire variable space it is easy to see that as the model increases in complexity the number of rules required to cover the variable space may become exceedingly large. This is particularly true when developing more detailed models of how to infer distances and directions for movement. The rule-based approach may be most useful to explore in the context of the process of landscape evaluation. If all animal objects are assumed to evaluate the landscape in exactly the same manner then landscape evaluation for either foraging, exploration, or dispersal purposes may be treated as a preprocessing stage thus reducing the computational burden of the simulation itself. Depending on how the rule-based model is structured, the spatially explicit nature of the problem may introduce further complexities in the rule-base. Nevertheless, it may provide valuable insight into how to develop fuzzy rule-bases that have spatially explicit outcomes that can be used in these ecological models. Furthermore, merely going through the exercise of constructing a plausible rule-based model may provide insights into the ecological problem and perhaps lead to the identification of gaps in our knowledge of a species that may be addressed by further ecological research.

4.2   Movement Direction and Memory

Although some small amount of movement memory was incorporated in the exploratory search strategy, it has been suggested dispersal models may benefit from incorporating more long term directionality memory [10]. This is something that may be accomplished by representing past directions as fuzzy concepts of direction, or even as fuzzy number. In the case of foraging and exploratory behavior some animals may display a directional bias towards the general center of their home range.

4.3   Fuzzy Logic and Robotics

Although these models do not typically endow animal objects with perceptual functions such as visual, auditory, or other sensors, they do impose a perceptual range that, in conjunction with retrieval of information from the GIS database, provides a similar role as do the perceptual functions of an intelligent robot. Similarly, like an intelligent robot, once the animal object has some intelligent information-processing functions that allow it to process the incoming information with respect to a task. One such task may be that of deciding which spatial object to move to as part of foraging behavior. Although there are no mechanical functions like those in a functioning robot, there are still software controls that allow the animal object to move and act as desired. The extent of similarity between these models and intelligent robots suggest that the body of research on using fuzzy logic to control robotic behavior [22] may provide general insights into problems of controlling the behavior of the individuals in these spatially explicit models.

4.4   Defining Fuzzy Spatial Relations

When incorporating fuzzy logic into any modeling system such as the spatially explicit ecological models discussed in this chapter, one of the major issues is how to arrive at the membership functions in such a manner that they are plausible. In ecological modeling there are two major potential sources upon which to base the functions. Ecological knowledge in the form of expert opinion, general observations, and the like are common. Thus, a knowledge acquisition approach similar that which has been used in the domain of soil mapping [26] could be adapted rather directly to be applied in the habitat evaluation component of this modeling system. It is clear that fuzzy spatial relations play an extremely important role in deriving an evaluation of how habitat is distributed over the landscape as well as important controls on animal object behavior. In this regard, the interactive human-machine approach [18] to acquire spatial relations could be modified with constraints such as ensuring transitivity to efficiently acquire representations of spatial relations by having a system interrogate the expert in that particular ecological domain. Secondly, field data can be used to arrive at plausible membership functions. It has rarely been attempted. However one study using a fuzzy rule-base model constructed a plausible model structure then altered it successively by modifying the number and shapes of the different fuzzy sets as well as the corresponding rules until a maximum correspondence with field data was achieved [12].

4.5   GIS Database Issues

Rather than rely on a spatial database that has been processed to the point of 'crispness', the explicit and pervasive incorporation of fuzzy sets in these models will allow a more direct use of spatial data that has been represented using a fuzzy spatial object model [e.g., 6,8]. In addition to the fuzziness of the spatial data being represented, it may also be exploited in the retrieval stage. This is one application area which did not receive explicit attention, but could be a useful exploitation of the inherent fuzziness of both attribute and spatial data. In a spatially explicit modeling study of sciurids, it was noted that the general approach used in that study provided a means of translating verbal or conceptual models to quantitative predictions [23]. It did not use fuzzy logic, thus was unable to formalize linguistic variables as is possible in fuzzy logic. Along those lines, a study in autonomous robotic navigation noted that the most peculiar feature of fuzzy logic is its intrinsic ability to integrate numeric (fuzzy) and symbolic (logic) aspects of reasoning [22]. This feature has not gone completely unnoticed by those working in ecological modeling. Two studies using fuzzy rule-base modeling approaches used knowledge based approaches to estimate quantitative results that were then compared to field observations. In both cases, a fuzzy model performed well [3,12].

4.6   Concluding Comment

Finally, perhaps the most exciting aspect of this modeling domain is its ability to integrate computational simulation modeling and GIS database issues with intelligent systems research but also have a direct interplay with ecological field-based research. For example, in a study of species density of foliage-dwelling spiders using a fuzzy rule-based model, the initial model structure was altered successively by modifying the number and shapes of the different fuzzy sets as well as the corresponding rules until a maximum correspondence with field data was achieved [23]. This is but one small example of how this modeling domain allows not only the simulation of reality-based behavior but also allows for a particularly productive interplay between simulation modeling and field ecology. The incorporation of fuzzy sets in these modeling efforts should enable the models to be more directly related to the results of field observations, hence more realistic.

References

1.   K. A. Baum and W. A. Grant. Hummingbird foraging behavior in different patch types: simulation of alternative strategies. *Ecological Modeling*, 137:201-209, 2001.

2.   J. C. Bezdek, R. Ehrlich, and W. Full. FCM: the fuzzy c-means clustering algorithm. *Computers and Geosciences*, 10:191-203, 1984.

3.   W. Bock and A. Salski. A fuzzy knowledge-based model of population dynamics of the yellow-necked mouse (Apodemus flavicollis) in a beech forest. *Ecological Modelling*, 108:155-161, 1998.

4.   M. A. Burgman, D. R. Breininger, B. W. Duncan, and S. Ferson. Setting reliability bounds on habitat suitability indices. *Ecological Applications*, 11(1):70-78, 2001.

5.   G. Cao. The definition of the niche by fuzzy set theory. *Ecological Modelling*, 77:65-71, 1995.

6.   T. Cheng, M. Molenaar, and H. Lin. Formalizing fuzzy objects from uncertain classification results. *International Journal Geographical Information Science*, 15(1):27-42, 2001.

7.   M. A. Cobb, F. E. Petry, and K. B. Shaw. Extensions to geometric approximations of spatial boundaries and assessment of fuzzy spatial relations. *International Journal of Fuzzy Sets and Systems*, 113(1):111-120, 2000.

8.   V. Cross and A. Firat. Fuzzy objects for geographical information systems. *International Journal of Fuzzy Sets and Systems*, 113(1):19-36, 2000.

9.   G. M. Foody. Fuzzy modelling of vegetation from remotely sensed imagery. *Ecological Modelling*, 85:3-12, 1996.

10. E. J. Gustafson and R. H. Gardner. The effect of landscape heterogeneity on the probability of patch colonization. *Ecology*, 77(1):94-107, 1996.

11. R. D. Holt, S. W. Pacala, T. W. Smith, and J. Liu. Linking contemporary vegetation models with spatially explicit animal population models. *Ecological Applications*, 5(1):20-27, 1995.

12. C. Kampichler, J. Barthel, and R. Wieland. Species density of foliage-dwelling spiders in field margins; a simple fuzzy rule-based model. *Ecological Modelling*, 129:87-99, 2000.

13. S. L. Lima and P. A. Zollner. Towards a behavioral ecology of ecological landscapes. *Trends in Ecology and Evolution*, 11(3):131-135, 1996.

14. A. Lomnicki. Individual-based models and the individual-based approach to population ecology. *Ecological Modelling*, 115:191-198, 1999.

15. D. Papadias, N. Karacapilidis, and D. Arkoumanis. Processing fuzzy spatial queries: a configuration similarity approach. *International Journal of Geographical Information Science*, 13(2):93-118, 1999.

16. S. F. Railsback, R. H. Lamberson, B. C. Harvey, and W. E. Duffy. Movement rules for individual-based models of stream fish. *Ecological Modelling*, 123:73-89, 1999.

17. B. W. Rickel, B. Anderson, and R. Pope. Using fuzzy systems, object-oriented programming, and GIS to evaluate wildlife habitat. *AI Applications*, 12(1/3):31-40, 1998.

18. V. B. Robinson. Individual and multipersonal fuzzy spatial relations acquired using human-machine interaction. *Fuzzy Sets and Systems*, 113(1):133-145, 2000.

19. V. B. Robinson. On fuzzy sets and the management of uncertainty in an intelligent geographic information system. In: G. Bordogna and G. Pasi (Eds.) *Recent Issues on Fuzzy Databases*, Physica-Verlag, pages 109-127, 2000.

20. V. B. Robinson and D. Thongs. Fuzzy set theory applied to the mixed pixel problem of multispectral landcover databases. In: B. K. Optiz (Ed.) *Geographic Information Systems in Government*, A. Deepak Publishing, Hampton, VA, pages 871-886, 1986.

21. M. Ruckelshaus, C. Hartway, and P. Kareiva. Assessing the data requirements of spatially explicit dispersal models. *Conservation Biology*, 11(6):1298-1306, 1997.

22. A. Saffioti. The uses of fuzzy logic in autonomous robot navigation: a catalogue raisonne. *Soft Computing*, 1(4):180-197, Springer-Verlag, 1997.

23. A. South. Extrapolating from individual movement behavior to population spacing patterns in a ranging mammal. *Ecological Modelling*, 117:343-360, 1999.

24. J. D. Westervelt and L. D. Hopkins. Modeling mobile individuals in dynamic landscapes. *Int. Journal of Geographical Information Science*, 13(3):191-208, 1999.

25. J. Zhang and N. Stuart. Fuzzy methods for categorical mapping with image-based land cover data. *International Journal of Geographical Information Science*, 15(2):175-195, 2001.

26. A. Zhu. A personal construct-based knowledge acquisition process for natural resource mapping. *International Journal of Geographical Information Science*, 13(2): 119-141, 1999.

27. P. A. Zollner. Comparing the landscape level perceptual abilities of forest sciruids in fragmented agricultural landscapes Landscape. *Ecology*, 15:523-533, 2000.

28. P. A. Zollner and S. L. Lima. Search strategies for landscape-level interpatch movements. *Ecology*, 80(3):1019-1030, 1999.

A Fuzzy Set Model of Approximate Linguistic Terms in Descriptions of Binary Topological Relations Between Simple Regions

F. Benjamin Zhan

Department of Geography
Southwest Texas State University
San Marcos, Texas 78666, USA
fbzhan@swt.edu

Abstract. While handling geospatial data, one often faces at least two types of fuzziness. The first type of fuzziness is found in the use of approximate linguistic terms to describe spatial objects and spatial relations. For instance, in "The flash flood in October of 1988 nearly completely flooded downtown San Marcos," the term "nearly completely" is approximate in nature. The second type of fuzziness is due to the indeterminate nature of the boundaries of some spatial objects. Good examples of such objects are climatic regions (e.g., hot versus warm regions) and polygons showing different soil types. This chapter is only concerned with the first type of fuzziness. It develops a fuzzy set model of approximate linguistic terms used in descriptions of binary topological relations between simple regions. After discussing related work, the author reviews cognitive evidences that demonstrate the fuzziness of approximate linguistic terms. Then a fuzzy set model of three approximate linguistic terms, 'a little bit,' 'somewhat,' and 'nearly completely,' is presented. A discussion of possible further research is followed by a summary at the end of the chapter.

Keywords. Geographic Information Science, spatial relations, approximate linguistic terms, fuzzy sets.

1 Introduction

People usually encounter at least two types of fuzziness while handling spatial data and processing spatial relations. The first type of fuzziness results from the approximate nature in the ways in which humans describe and process spatial data and spatial relations. The second type of fuzziness is due to the indeterminate nature in the boundaries of some spatial objects. In the literature of Geographic Information Science (GIScience), a spatial object is defined as the computerized representation of a geographic entity that cannot be further divided into its own type (e.g., a lake). Spatial relations are relations between spatial objects, including distance, directional, and topological relations.

One example may help illustrate the existence of the first type of fuzziness. For example, people often tend to use approximate terms when they describe spatial relations. When describing topological relations between two regions, they may use one of the four statements given below to describe the degree of "Region Q covering Region R." The main differences in the four statements are the different approximate terms used. These approximate terms are called *approximate linguistic terms*.

(1) Region Q covers region R *a little bit*;
(2) Region Q covers region R *somewhat*;
(3) Region Q *nearly completely* covers region R;
(4) Region Q *completely* covers region R.

In daily communications, humans have little difficulty, if any, in understanding each other as to "how much" is "a little bit" in these statements. This seemingly very intuitive and simple task can be difficult for GIS (Geographic Information Systems) to handle because current GIS are not *intelligent* enough to directly represent and process spatial relations described in natural languages. For example, it is still a fairly difficult task for current GIS to directly process the query given below.

"Find all areas in central Texas that will be *nearly completely* flooded (covered) by the Colorado River when the water level of the river reaches H meters."

The reason that current GIS cannot process this type of query directly is because they do not have a mechanism in the system to interpret how much "nearly completely" is. In order to accommodate this type of query in GIS, we need to first understand how the human mind processes natural language descriptions of topological relations containing approximate linguistic terms. We then may be able to develop adequate mathematical and computational models that can be used to accommodate queries containing approximate linguistic terms typically used by humans in their daily communications in a GIS environment.

The second type of fuzziness exists in many areas of spatial data handling. In both socioeconomic and natural resource data, spatial objects tend to have indeterminate boundaries [4]. Spatial objects with indeterminate boundaries are usually called *fuzzy* spatial objects [52]. Typical examples of fuzzy spatial objects are climatological regions [24] and polygons depicting different categories of soils [3]. Not all spatial objects have an indeterminate boundary. Some spatial objects are crisp objects because their boundaries are clearly defined. Good examples of crisp spatial objects are home sites whose boundaries are clearly defined through high precision survey. Almost all commercially available GIS packages today employ data models that only handle crisp spatial objects.

This chapter is only concerned with the first type of fuzziness involving spatial objects. The goal of this study is to develop a fuzzy set [51] model of approximate linguistic terms that can be used to model the fuzziness exhibited in the descriptions of binary topological relations between simple crisp regions. A simple crisp region in this context is a convex crisp region without holes. We will focus this discussion on the topological relation "cover" between simple regions.

The discussion that follows begins with a brief review of the literature of spatial relations and the cognitive aspects of spatial relations. We will then discuss some preliminary cognitive evidences illustrating the fuzziness related to approximate linguistic terms in descriptions of topological relations. Based on these cognitive evidences, a fuzzy set model of approximate linguistic terms is developed and presented. A discussion of issues subject to future research precedes a summary of this discussion at the end of this chapter.

2   Related Literature

The development of general theories of spatial relations has been of interest to scholars from various disciplines for more than two decades [18,1,34,36,32,12, 26]. So far significant advancements have been made in the development of mathematical and computational models of distance and directional spatial relations [35,21,31], mathematical and computational models of topological relations [10,19,5,12,6], and the cognitive aspects of spatial relations [27,28,25, 29,30]. It is not the intention of this discussion to provide an exhaustive literature review on these topics. Instead, we will briefly review previous work that is directly related to the materials presented in this chapter.

2.1   The 9-Intersection Model of Topological Relations

A notable advancement in searching for mathematical models of spatial relations is the development of the 9-Intersection model of binary topological relations between crisp spatial objects [12]. In this model, a spatial object is considered to consist of three parts: an *interior*, a *boundary,* and an *exterior*. Topological relations between two crisp objects A and B can be determined by the values of the intersections of these three parts of the two objects. There are a total of nine intersections among the six parts of the two objects. Depending on the values of the intersections, which can either be empty or non-empty, a total of 512 (2^9) possible combinations of topological relations can be distinguished between two spatial objects.

In a two-dimensional space, eight binary topological relations between simple spatial regions without holes, 18 binary topological relations between regions with holes, 33 binary relations between two simple lines, and 19 between a simple line and a region without holes can be distinguished [12]. The eight binary topological relations and their prototypical examples are given in Figure 1. These eight topological relations are "disjoint," "meet," "covers," "coveredby," "contain," "inside," "equal," and "overlap" (Fig. 1). No names have been given to the other prototypical topological relations. Figure 2 shows prototypical examples of the 19 binary topological relations. These prototypical examples provide very good frameworks within which further studies of topological relations can be performed. It should be noted that the names for the eight topological relations between simple crisp regions are informational only in this discussion. The term 'covers' in the eight topological relations is different from that of the term 'cover' used in the four statements given in Section 1 (see Section 2.4 for more discussions on this issue).

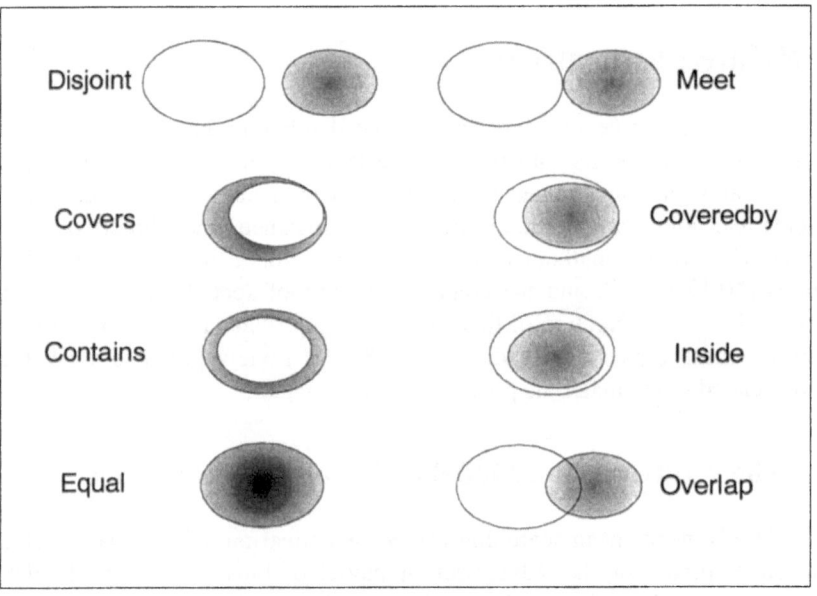

Fig. 1. Eight binary topological relations between simple regions without holes as distinguished by the "9-Intersection" model proposed by Egenhofer and Herring [12].

2.2   Cognitive Aspects of Spatial Relations

The literature on the cognitive aspect of geographic space and cognitive models of spatial relations is extensive (see Mark [26]; Mark et al. [30]; and the references therein). While searching for cognitive models of spatial relations, the fundamental

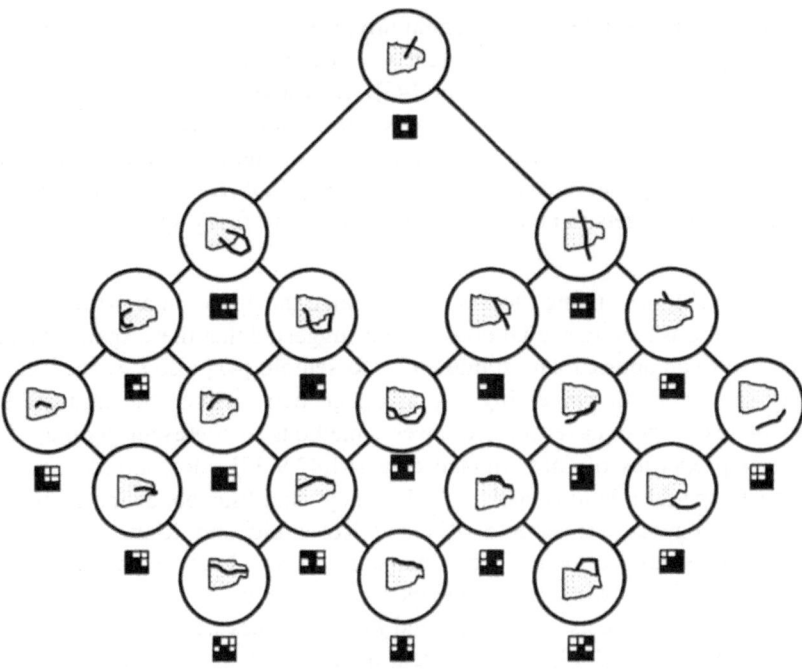

Fig. 2. The 19 binary topological relations between a linear feature and a region as distinguished by the "9-Intersection" model proposed by Egenhofer and Herring. (After Mark and Egenhofer [27].)

questions that need to be addressed are how humans learn, perceive, and represent spatial relations in their mind, and how cognitive models of spatial relations, once developed, can be used to help us develop better computerized models so that we can design better computer systems to process spatial relations in GIS in a more natural way.

Based on prototypical examples of the 9-Intersection model, Mark, Egenhofer, and their colleagues have tested whether spatial relations in two-dimensional space distinguished by the 9-Intersection model are confirmed by different natural languages. Their findings suggest that the 9-Intersection model is a valid model, but humans tend to use more aggregated categories of the topological relations than those distinguished by the 9-Intersection model [27,28,13,26].

2.3   Models of Spatial Relations Between Fuzzy Regions

The last two subsections were devoted to discussions about mathematical models and cognitive aspects of spatial relations between crisp spatial objects. We now turn to review previous work related to models of spatial relations between fuzzy spatial objects. In the GISscience community, it is well recognized that fuzziness exhibits in geographic data and current GIS cannot be used to process fuzzy geographic data directly (see Unwin [48], Burrough and Frank [4], and the references therein). In fact, there have been considerable research efforts concerned with the development of methods for handling fuzzy geographic data in the literature of spatial analysis [23,24], GIS [37,38,15-17,50,2,49], and the application of fuzzy sets in modeling fuzzy objects and fuzzy spatial relations [38-40,4,14,52,53].

Freeman [18] was among the first to recognize the approximate (fuzzy) nature of spatial relations among spatial objects, and suggested that these spatial relations be described in an approximate framework. Rosenfeld [41] also saw the potential of fuzzy set theory in processing digital image data. Since the mid 1970s, a number of researchers have defined several methods to represent geometric and topological properties of fuzzy image data [41,42,9,33]. Rosenfeld and Klette [43] defined spatial relations such as "adjacency" and "surroundedness" between fuzzy image regions. Rosenfeld [41] and Rosenfeld and Klette [43] defined spatial relationships between fuzzy image regions. Dubois and Jaulent [9] extended Rosenfeld's work and developed a general model of spatial relations between two fuzzy objects. Robinson and his colleagues tried to quantify "near" in distance relations [38-40]. More recent development in this area includes the development of methods that can be used to support some operations on fuzzy regions (Erwig and Schneider 1997) and models of topological relations between fuzzy regions [53].

In geographical analysis, Leung [23,24] developed methods for representing fuzzy regions in a geographical context and discussed the treatment of some fundamental concepts of spatial analysis using fuzzy set theory. These concepts include distances, directions, and regions used in geographical analysis. In image analysis, Krishnapuram et al. [21] and Matsakis et al. [31] developed methods for quantitatively analyzing the directional relations between fuzzy regions. More recently, Clementini and DiFelice [5-7] and Cohn and Gotts [8] have developed models of topological relations between fuzzy regions from different perspectives. Their models, however, treat the indeterminate boundary of a fuzzy region as a *thick* boundary. As a consequence, finer distinctions between points (pixels) lying within the *thick* boundary cannot be made in these models.

Zhan [52,53] proposed an approximate model of binary topological relations between simple fuzzy regions using fuzzy set theory. In this model, a fuzzy region is considered to have three parts, the *core*, the *indeterminate boundary*, and the *exterior* (Fig. 3). Membership values within the core are always 1.0. Membership values in the indeterminate boundary change from 1.0 to 0.0 from the inside edge to the outside edge of the indeterminate boundary. Membership values in the exterior are always 0.

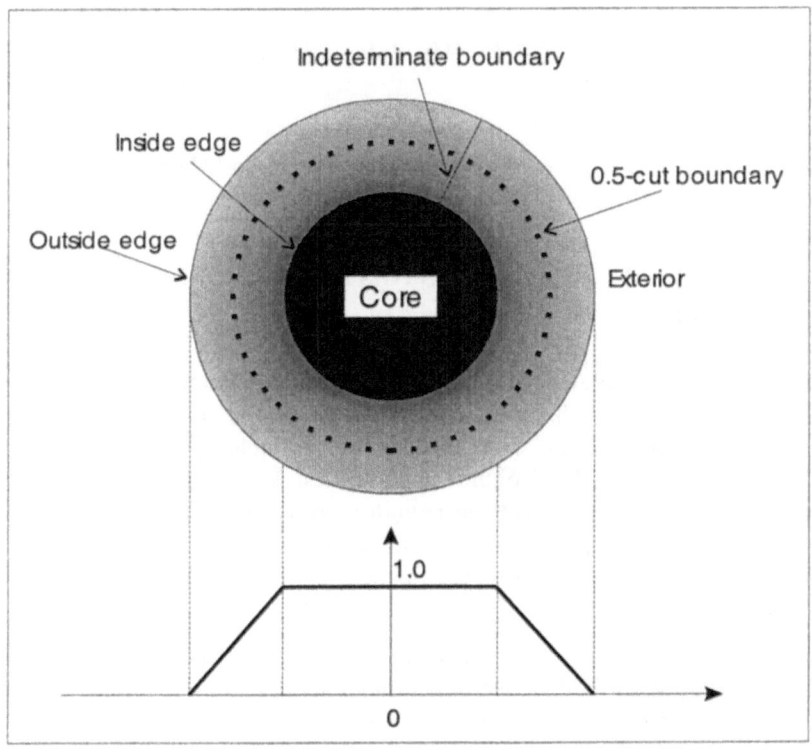

Fig. 3. A simple fuzzy region and its components.

Using the definition of a fuzzy region given above, and based on the concept of $\alpha-$ cuts and fuzzy sets [20], Zhan [52,53] proposed a method for computing the fuzzy membership values of a given binary topological relation between fuzzy regions in relation to the eight prototypical topological relations between two regions in two dimensional space as suggested by the 9-Intersection model. The general formula for computing the fuzzy membership value ($\tau(A,B)$) for a given prototypical binary topological relation between two simple fuzzy regions is given in (1). An example formula for determining the membership value for relation 'overlap' is given in (2). In these expressions, α_i is the fuzzy membership value used to generate α-cut region $A_{\alpha i}$, and $\tau(A_{\alpha i}, B_{\alpha j})$ is the topological relation between α-cut regions $A_{\alpha i}$ and $A_{\alpha j}$. Any α-cut region is a crisp region. The utilization of α-cut regions transforms the computation of fuzzy membership values of binary topological relations between two fuzzy regions into computation of aggregated binary topological relations between crisp regions, and thus makes it possible to determine topological relations between fuzzy regions within a current GIS environment.

$$\tau(A,B) = \sum_{i=1}^{n} \sum_{j=1}^{n} (\alpha_i - \alpha_{i+1})(\alpha_j - \alpha_{j+1}) \tau(A_{\alpha i}, B_{\alpha j}) \qquad (1)$$

$$\tau_{overlap}(A,B) = \sum_{i=1}^{n} \sum_{j=1}^{n} (\alpha_i - \alpha_{i+1})(\alpha_j - \alpha_{j+1}) \tau_{overlap}(A_{\alpha i}, B_{\alpha j}) \qquad (2)$$

2.4 Approximate Linguistic Terms in Descriptions of Spatial Relations

The significance of linguistic descriptions of spatial relations and the roles of linguistic descriptions in helping us understand how humans process spatial relations have long been recognized by researchers in various areas [45,25,29,46,47,22,44]. But to the author's best knowledge, there has been limited published research about the cognitive aspect of approximate linguistic terms in descriptions of spatial relations and the development of adequate models that can be used to represent approximate linguistic terms in a computing environment.

Approximate linguistic terms, however, are present in many natural language descriptions of spatial relations in similar situations as mentioned in the examples given at the beginning of this chapter. In the eight topological relations between simple crisp regions as distinguished by the 9-Intersection model, approximate linguistic terms may be present in relations 'overlap,' 'cover,' and 'coveredby.' It is debatable whether 'overlap' and 'cover' are interchangeable in some situations. People in their daily communications use statements like the four given in the introduction section naturally. But the meaning of the word, cover, in those four statements is not the same as the one suggested by the 9-Intersection model. For this discussion, We will follow the natural intuition reflected in the four statements given at the introduction section, and use what 'cover' means in those statements. In these four statements, 'Region Q covers Region R' means Region Q may 'embrace' part or the whole of Region R. According to the 9-Intersection model, the sentence, 'Region Q covers Region R,' suggests that Region Q completely embraces Region R *and* the boundaries of both regions also meet.

Approximate linguistic terms can also be used in natural language descriptions of topological relations between a linear feature (e.g., a road) and a region (e.g., a park). For instance, one may say that: This road goes *very deep* into the park; this road only extends to the park *a little*; this road *almost* crosses the park. One may find similar examples in other descriptions. A better understanding of the cognitive aspects of these approximate linguistic terms and the development of adequate mathematical and computational models to represent these terms are necessary for developing more intelligent GIS.

3 Fuzziness of Approximate Linguistic Terms — Preliminary Cognitive Evidences

In order to obtain cognitive evidences regarding the fuzzy nature of approximate linguistic terms in the descriptions of binary topological relations between simple crisp regions, the author designed two experiments and tested those approximate linguistic terms and their corresponding topological relations through human subject tests [54]. We will briefly review the experiments and the results in this section.

3.1 Experimental Design

In Experiment One, subjects were given ten different diagrams showing varying degrees of region Q covering region R. Region Q is represented by a circle and Region R is depicted by an ellipse. Associated with each diagram is a set of four statements. Each diagram can be described using one of the four statements (Fig. 4). Written instructions were provided to the subjects who participated in the experiment (see below). The diagrams were deliberately arranged so that the progression of the degree of region Q covering region R in the ten diagrams is not the same sequence in which the ten diagrams appeared in the experiment.

> **Instructions to participants of Experiment One**: *"On the next two pages, there are ten diagrams showing different degrees of 'Region Q covers Region R.' For each diagram, there are four statements next to it. Please choose one and only one statement from the four statements that best describes the degree of 'Region Q covers Region R' in each diagram."*

In the second experiment (Experiment Two), subjects were given an example of Region Q (a circle) and an example of Region R (an ellipse) as well as the four statements. The subjects were then asked to draw diagrams illustrating each of the four statements. Given below are the instructions presented to subjects who participated in Experiment Two. One set of example drawings from a subject who participated in Experiment Two is shown in Figure 5.

> **Instructions to participants of Experiment Two**: *"There are two regions, Q and R, as shown below. Please draw a diagram that best matches each of the statements given below."*

Responses from both experiments were analyzed using an area-based ratio. An area-based ratio is the ratio between the area of the overlapping portion of the two regions and the area of Region R in the case of Region Q covering Region R. More specifically, in the spatial relation of region Q covering region R,

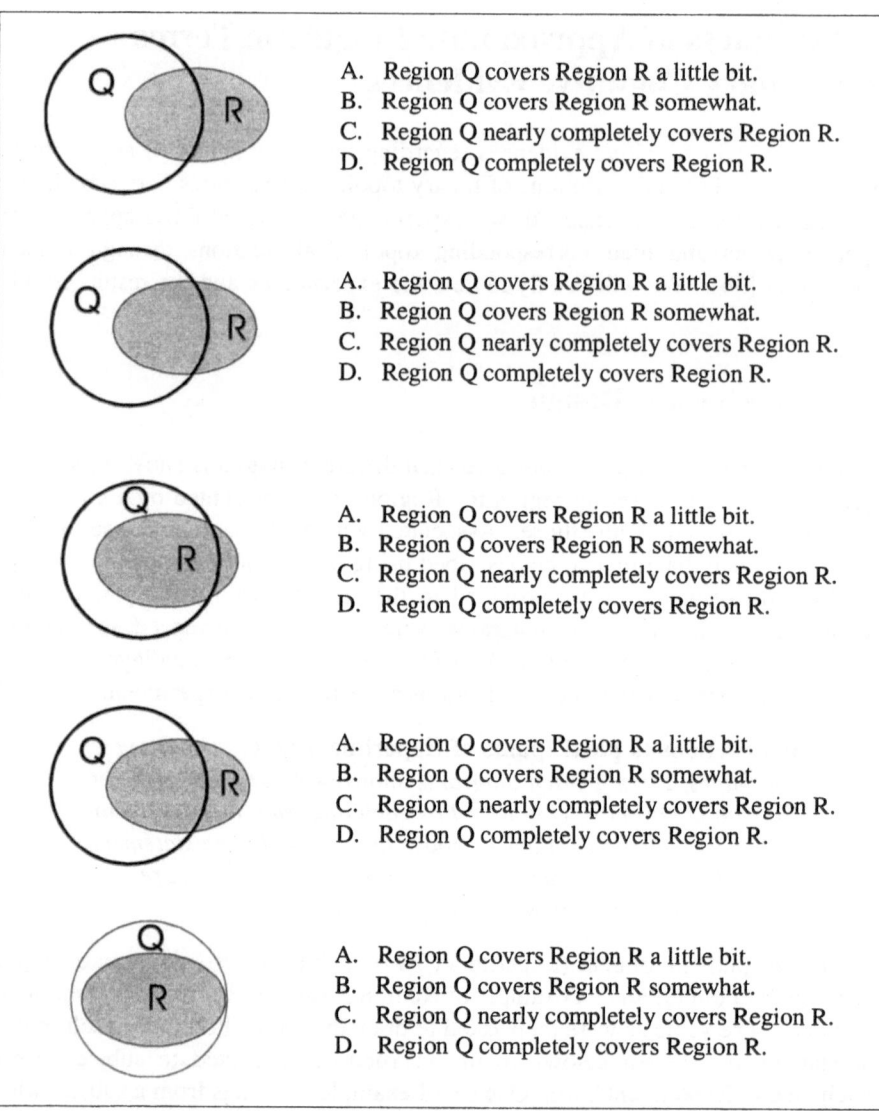

Fig. 4. Examples of stimuli used in Experiment One.

three areas can be distinguished, the area of Region Q (denoted as A_Q), the area of Region R (denoted as A_R), and the area covered by the overlapping portion of Regions Q and R (denoted as A_O). In the case of Region Q covering Region R, the area-based ratio (R_A) is defined as the ratio between A_O and A_R.

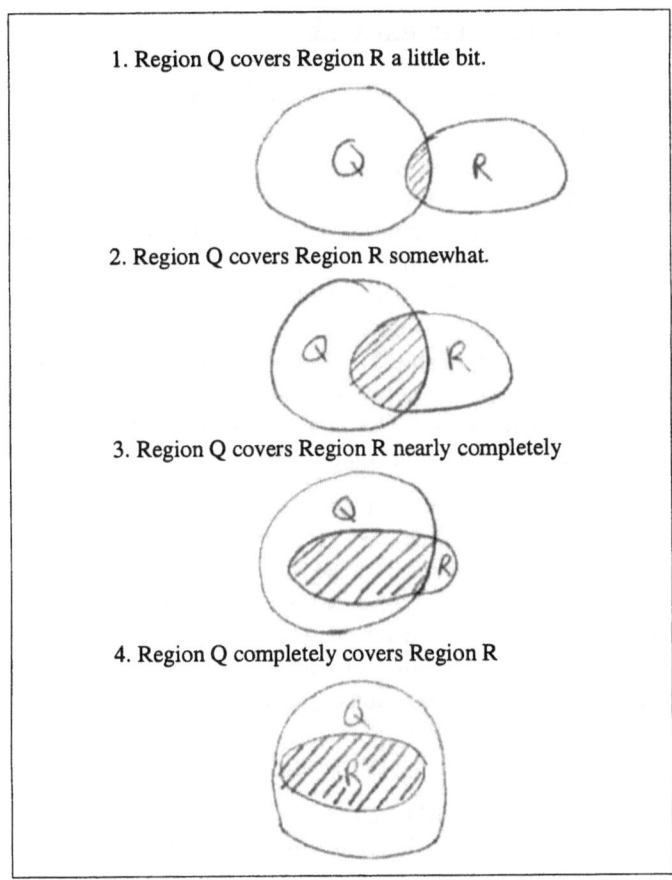

1. Region Q covers Region R a little bit.

2. Region Q covers Region R somewhat.

3. Region Q covers Region R nearly completely

4. Region Q completely covers Region R

Fig. 5. Example drawings by a subject in Experiment Two.

The task of analyzing the responses in Experiment One then can be accomplished in two steps. In the first step, an area-based ratio for each of the ten stimuli is computed. In the second step, these calculated area-based ratios are assigned to the corresponding choices of the subjects from the four statements. For Experiment Two, the analysis task is to calculate the area-based ratio corresponding to each of the drawings produced by the subjects. Stimuli used in Experiment One and all drawings in Experiment Two were scanned and saved as TIFF image files. These image files were then imported to the ArcView GIS package in order to measure the A_R and A_O for each case and relate them to each corresponding stimulus or drawing.

3.2   Results from Experiment One

Results from the responses of 41 subjects who participated in Experiment One are summarized in Table 1. For each of the four statements, the number and percentage of subjects who chose a particular statement for a specific diagram were computed (Table 1). The resulting area-based ratios corresponding to each of the diagrams are given in the rightmost column of Table 1. As can be seen in Table 1, nearly all subjects chose the first statement of "Region Q covers Region R a little bit" for diagrams 1 and 2. The area-based ratios for diagrams 1 and 2 are 0.01 and 0.08. It is interesting to notice that some subjects also picked up this statement for diagrams 3 to 6 (Table 1). Among these subjects, it is particularly interesting, and somewhat surprising, to note that 31 out the 41 subjects selected the first statement for diagram 4. The area-based ratio related to diagram 4 is 0.27, which is much larger than that of diagram 1.

The distribution of the number of subjects who selected the statement of "Region Q covers Region R somewhat" for one of the ten diagrams is more complex compared to the distribution of the number of subjects who selected the statement of "Region Q covers Region R a little bit" for one of the ten diagrams. Most subjects (39 out of 41) picked up this statement for diagram 6. Diagram 6 has an area-based ratio of 0.50. In addition, 83% of the subjects selected this statement for diagrams 3 and 5. No subject picked up the statement of "Region Q covers Region R somewhat" for diagrams 1 and 10. This result is to be expected because these two diagrams are at the two ends of the spectrum (Table 1).

Most subjects (40 out of 41) selected the statement "Region Q nearly completely covers Region R" for diagram 9 that has an area-based ratio of 0.90 (Table 1). In addition, some subjects also selected this statement for diagrams 5, 7, 8, and 10. It was a fairly easy task for most of the subjects (40 of 41) to identify diagram 10 as the diagram showing "Region Q completely covers Region R." One subject selected this statement for diagram 1, which apparently was an error.

Results from Experiment One suggest that: (1) if the area-based ratio is less than 0.08, then nearly everyone agrees that "Region Q covers Region R a little bit;" (2) if the ratio is between 0.15 and 0.50, then some people may still consider that "Region Q covers Region R a little bit;" (3) in the case of "Region Q covering Region R somewhat," the area-based ratio can vary in a wide range from 0.15 to 0.70, and the majority of the participants in the experiment seem to consider an area-based ratio of 0.50 as "somewhat;" (4) if the area-based ratio is greater than or equal to 0.65 but less than 0.99, then one usually says that "Region Q nearly completely covers Region R," and the majority of subjects seem to consider an area-based ratio of 0.90 as "nearly completely" cover; (5) if the area-based ratio is 1.00, then there is little doubt that one usually says that "Region Q completely covers Region R." These results are summarized in Table 2.

Table 1. Results of Experiment One: Distribution of subjects' choices of the four statements related to each of the ten stimuli. (These four statements are: (1) Region Q covers region R *a little bit*; (2) Region Q covers region R *somewhat*; (3) Region Q *nearly completely* covers region R; (4) Region Q *completely* covers region R.)

Stimuli		Number of subjects chose the description				Percentage of subjects chose the description				Area-based ratio of Q covering R in the stimuli
ID	Diagram	A	B	C	D	A	B	C	D	
1		40	1			98%	2%			0.01
2		41				100%				**0.08**
3		7	34			17%	83%			0.15
4		31	10			76%	24%			0.27
5		6	34	1		15%	83%	2%		0.39
6		2	39			5%	95%			**0.50**
7			27	14			66%	34%		0.65
8			18	22	1		44%	54%	2%	0.70
9			1	40			2%	98%		**0.90**
10		1		1	39	2%		2%	95%	**0.99**

Table 2. Correspondance between approximate linguistic terms and area-based ratios based on results from Experiment One.

Approximate linguistic terms	Area-based ratio (R_A)
a little bit	$R_A <= 0.50$
somewhat	$0.15 <= R_A <= 0.70$
nearly completely	$0.65 <= R_A <= 0.99$
completely	$R_A = 1.00$

3.3   Results from Experiment Two

The remaining question is to test whether the area-based ratios corresponding to different approximate linguistic terms selected by the majority of the subjects in Experiment One are confirmed by the results from Experiment Two. This question can be answered through the computation of the mean of the area-based ratios for drawings from the subjects who participated in Experiment Two. The author obtained a total of 32 useful responses from Experiment Two. Area-based ratios related to these 32 responses were calculated and they are presented in Table 3. All of the drawings corresponding to the fourth statement, "Region Q nearly completely covers Region R," had Region R completely within Region Q. Therefore, the area-based ratios for all drawings related to the fourth statement were all 1.00, and hence are not reported in Table 3.

The mean area-based ratio related to the first statement, "Region Q covers Region R a little bit," is 0.10 with a standard deviation of 0.07; the mean ratio related to the second statement, "Region Q covers Region R somewhat," is 0.36 with a standard deviation of 0.10; the mean ratio related to the third statement, "Region Q nearly completely covers Region R," is 0.74 with a standard deviation of 0.18 (Table 3). The minimum area-based ratios for the first, second, and third statement are 0.02, 0.13, and 0.25. The maximum area-based ratios for the first, second, and third statement are 0.29, 0.55, and 1.00, respectively (Table 3).

The results given above are fairly consistent with the results obtained from Experiment One. Based on the mean, the standard deviation, and the minimum and maximum values of the area-based ratios, the results from Experiment Two suggest that: (1) for the approximate linguistic term "a little bit," the mean of the area-based ratios from the drawings is 0.10 with a standard deviation of 0.07; (2) for the term "somewhat," the mean of the area-based ratios is 0.36 with a standard deviation of 0.10; (3) for the term "nearly completely," the mean is 0.74 with a standard deviation of 0.18; (4) if the area-based ratio is 1.00, then one can say "Region Q completely covers Region R."

The range of area-based ratios corresponding to different approximate linguistic terms discussed above changes somewhat from the ones obtained in Experiment One when the area-based ratios are allowed to vary two times of standard deviation from the mean (Tables 2 and 4).

A further comparison can be made between the largest percentage of subjects who chose a particular statement for a stimulus with a given area-based ratio in Experiment One and the mean of the area-based ratios in drawings corresponding to one of the three linguistic terms in Experiment Two. In Experiment One, the largest percentage of subjects who chose the first statement was for the stimulus with an area-based ratio of 0.08. This result is very close to the mean (0.10) of area-based ratios of drawings corresponding to the linguistic term 'a little bit' in

Table 3. Results of Experiment Two: Area-based ratios corresponding to the drawings from the subjects (A_{Ri}, - area covered by Region R in the drawings corresponding to the ith statement; A_{Oi} - area covered by the overlapping portion of Regions Q and R in the drawings corresponding to the ith statement; R_{Ai} – area-based ratio in the drawings corresponding to the ith statement).

Subject ID	A_{R1}	A_{O1}	R_{A1}	A_{R2}	A_{O2}	R_{A2}	A_{R3}	A_{O3}	R_{A3}
1	15510.5	1790.2	0.12	22687.8	10927.5	0.48	14698.3	12114.2	0.82
2	27373.1	2289.4	0.08	29005.7	12615.5	0.43	22867.8	17450.1	0.76
3	10162.4	585.3	0.06	10554.9	1414.2	0.13	7943.0	6383.4	0.80
4	7500.4	1245.3	0.17	9782.7	3435.9	0.35	7906.3	6354.6	0.80
5	32243.2	2724.9	0.08	22243.2	4990.2	0.22	13443.8	10975.6	0.82
6	25755.0	1258.4	0.05	22295.5	6293.1	0.28	27421.9	24525.5	0.89
7	3121.3	78.1	0.03	3986.8	1433.2	0.36	7697.3	3238.9	0.42
8	3594.9	279.5	0.08	4015.3	1574.5	0.39	2281.4	2284.4	1.00
9	1279.7	63.3	0.05	1047.9	259.9	0.25	638.4	561.9	0.88
10	1464.0	419.6	0.29	849.7	420.5	0.49	1901.8	1477.7	0.78
11	1681.1	356.5	0.21	1950.0	1021.2	0.52	2012.5	1663.9	0.83
12	1390.2	358.1	0.26	1527.1	582.7	0.38	1202.4	970.2	0.81
13	1839.3	108.7	0.06	3003.1	707.7	0.24	2380.0	1625.4	0.68
14	2651.3	181.3	0.07	1054.1	423.0	0.40	879.5	220.4	0.25
15	2115.1	152.1	0.07	2243.0	549.4	0.24	3058.6	2074.2	0.68
16	2490.7	179.3	0.07	1404.7	449.3	0.32	1971.6	1697.1	0.86
17	777.0	110.3	0.14	1174.7	271.8	0.23	1208.7	1078.6	0.89
18	3114.1	252.6	0.08	3406.7	943.1	0.28	2007.1	858.3	0.43
19	3240.6	251.1	0.08	2364.9	1192.7	0.50	2142.4	1673.3	0.78
20	1090.4	74.1	0.07	1841.7	497.6	0.27	1123.5	683.2	0.61
21	3446.9	138.8	0.04	3615.8	1978.2	0.55	3278.6	3009.9	0.92
22	2353.7	485.0	0.21	1909.5	977.9	0.51	1909.5	977.9	0.51
23	584.8	47.0	0.08	584.7	248.4	0.42	522.0	326.4	0.63
24	2561.7	169.6	0.07	1876.8	604.4	0.32	1689.8	1478.0	0.87
25	2936.5	627.2	0.21	3302.8	1195.7	0.36	3652.8	2303.2	0.63
26	3718.2	200.0	0.05	3498.2	1707.9	0.49	4037.2	3638.8	0.90
27	1196.5	167.7	0.14	1050.9	346.0	0.33	909.8	837.9	0.92
28	2871.1	197.4	0.07	2158.6	705.6	0.33	2283.2	1869.9	0.82
29	3077.7	71.9	0.02	2820.3	844.1	0.30	2827.9	2008.7	0.71
30	3716.7	287.6	0.08	2275.7	1116.2	0.49	2896.4	2466.4	0.85
31	3035.9	223.0	0.07	2218.8	628.8	0.28	1630.5	1039.0	0.64
32	425.7	31.4	0.07	526.4	138.5	0.26	889.3	498.3	0.56
Mean			0.10			0.36			0.74
STDEV			0.07			0.10			0.18
Min/Max			0.02/0.29			0.13/0.55			0.25/1.00

Table 4. Correspondance between approximate linguistic terms and area-based ratios based on results from Experiment Two when the area-based ratios are allowed to vary between two times of standard deviation from the mean.

Approximate linguistic term	Area-based ratio (R_A)
a little bit	$R_A < 0.24$
somewhat	$0.16 <= R_A <= 0.56$
nearly completely	$0.38 <= R_A < 1.00$
completely	$R_A = 1.00$

Experiment Two. It is somewhat puzzling, however, that this consistency was not observed in results from both experiments regarding linguistic terms 'somewhat' and 'nearly completely.' For these two terms, the area-based ratios received the highest number of responses are 0.50 and 0.90 in Experiment One, but the means of area-based ratios for these two terms are 0.36 and 0.74 in Experiment Two. These discrepancies were a bit larger than expected. More tests are therefore needed in order to determine which ratio we need to adhere to.

4   A Fuzzy Set Model of Approximate Linguistic Terms

Based on discussions in the previous section, it is now possible to construct a fuzzy set model of the three approximate linguistic terms: a little bit, somewhat, and nearly completely, in relation to the topological relation described by 'Region Q covers Region R'. If we assume that fuzzy membership values change linearly within the interval of area-based ratios corresponding to a particular approximate linguistic term, then the membership functions can be constructed as shown in Figure 6. The fuzzy membership functions corresponding to the three approximate linguistic terms can be defined on three fuzzy sets. Each fuzzy set consists of the area-based ratios within the interval of area-based ratios related to each of the three approximate linguistic terms. Denote $V1(x)$, $V2(x)$, and $V3(x)$ as the three fuzzy membership functions corresponding to "a little bit," "somewhat," and "nearly completely." The fuzzy membership functions can be defined on the interval of $(0.0, 1.0)$ as shown in (3) to (5).

$$V1(x) = \begin{cases} 1 & when\ 0 < x \le 0.10 \\ (0.5 - x)/0.4 & when\ 0.10 < x \le 0.50 \\ 0 & when\ x > 0.50 \end{cases} \qquad (3)$$

$$V2(x) = \begin{cases} (x-0.15)/0.21 & when\ 0.15 < x \le 0.36 \\ (0.70-x)/0.34 & when\ 0.36 < x \le 0.70 \end{cases} \quad (4)$$

$$V3(x) = \begin{cases} 0 & when\ x < 0.65 \\ (x-0.65)/0.09 & when\ 0.65 < x \le 0.74 \\ 1 & when\ x > 0.74 \end{cases} \quad (5)$$

Based on these fuzzy membership functions, it is possible to determine the fuzzy membership value of any area-based ratios based on α-cuts [20]. Therefore, once the area-based ratio of an approximate linguistic term is given, the fuzzy membership value related to that area-based ratio is also determined based on (3) to (5).

Fuzzy membership functions described above were based on the assumption that membership values change linearly over the interval of area-based ratios in which membership values corresponding to a linguistic term are fuzzy. However, the cognitive evidences discussed in the last section do not completely support this assumption. Because there are an infinite number of area-based ratios for each approximate linguistic term, it is not possible to exhaustively test all possible area-based ratios and obtain their corresponding membership values through human

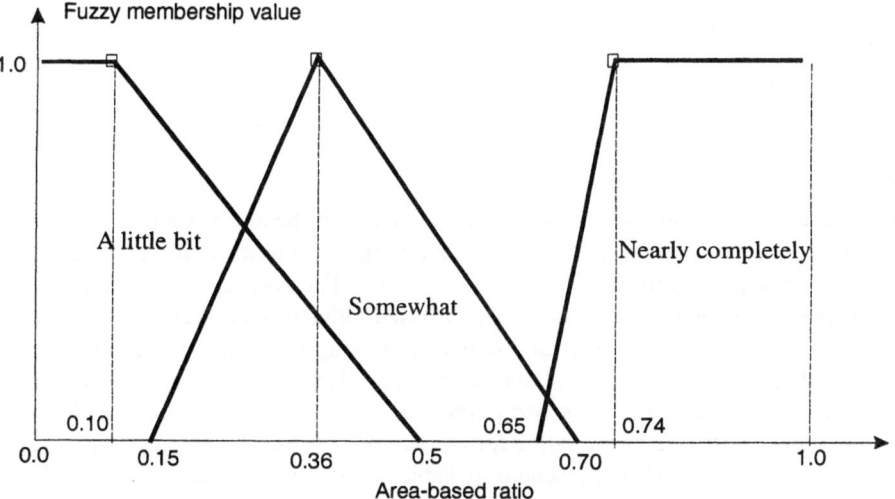

Fig. 6. Fuzzy membership values of approximate linguistic terms "a little bit," "somewhat," and "nearly completely."

subject tests. However, it is possible to construct preliminary membership functions based on the percentage of subjects who chose a particular statement for a given stimulus based on results from Experiment One (Table 1) in combination with the mean values of the area-based ratios obtained in Experiment Two (Table 3). These membership values are summarized in Table 5. Of course, a similar table can be constructed based on results from Experiment Two. But, membership values in Table 5 should suffice to illustrate the general idea that membership values do not necessarily change linearly over the interval of area-based ratios within which the membership values are fuzzy.

Table 5. Evidences of non-linear fuzzy membership functions.

Area-based ratio	Fuzzy membership value		
	A little bit	Somewhat	Nearly completely
0.01	0.98	0.02	0
0.08	1.00	0.0	0
0.15	0.17	0.83	0
0.27	0.76	0.24	0
0.39	0.15	0.83	0.02
0.50	0.05	0.95	0
0.65	0	0.66	0.34
0.70	0	0.44	0.54
0.90	0	0.02	0.98
0.99	0.02	0	0.02

5   Discussion

There are several extensions to the study presented in Sections 3 and 4. The first extension is to expand the list of approximate linguistic terms that may be used in natural language descriptions of spatial relations. The second extension is to test whether human cognition is based on area-based ratios or distance-based ratios. A distance-ratio is defined as the ratio between the length of the longest diameter of A_O and the length of the longest diameter of A_R. Here diameter is generalized to mean a straight line connecting two points on the boundary of A_O or A_R. Table 6 also gives the distance-based ratios for the ten stimuli used in Experiment One in addition to the information presented in Table 1. It is clear from the values of the distance-based ratios in Table 6 that the relationships between fuzzy membership values, distance-based ratios, and area-based ratios are consistent for regularly shaped convex regions without holes such as a circle or an ellipse. But distance-based ratios may introduce a distortion when irregularly shaped non-convex regions are involved.

Table 6. Distance-based versus area-based ratios. Distribution of subjects' choices of the four statements related to each of the ten diagrams. (These four statements are: (1) Region Q covers region R *a little bit*; (2) Region Q covers region R *somewhat*; (3) Region Q *nearly completely* covers region R; (4) Region Q *completely* covers region R.)

Stimuli		Number of subjects chose the description				Percentage of subjects chose the description				Distance/Area based ratio of 'Q covering R' in the stimuli
ID	Diagram	A	B	C	D	A	B	C	D	
1	(Q)(R)	40	1			98%	2%			0.04/0.01
2	(Q)(R)	41				100%				**0.15/0.08**
3	(Q)(R)	7	34			17%	83%			0.19/0.15
4	(Q)(R)	31	10			76%	24%			0.24/0.27
5	(Q)(R)	6	34	1		15%	83%	2%		0.38/0.39
6	(Q)(R)	2	39			5%	95%			**0.44/0.50**
7	(Q)(R)		27	14			66%	34%		0.65/0.65
8	(Q)(R)		18	22	1		44%	54%	2%	0.71/0.70
9	(Q)(R)			1	40		2%	98%		**0.87/0.90**
10	(Q)(R)	1		1	39	2%		2%	95%	**1.00/0.99**

A third extension along this line of work is to develop fuzzy set models for irregularly shaped regions including both convex and non-convex regions with holes. Figure 7 lists ten polygons of different shapes. These polygons are census block group polygons extracted from census geographic area files for Travis County in Texas. Future human subject tests and fuzzy set model development will definitely need to involve real world irregularly shaped convex and non-convex polygons with or without holes.

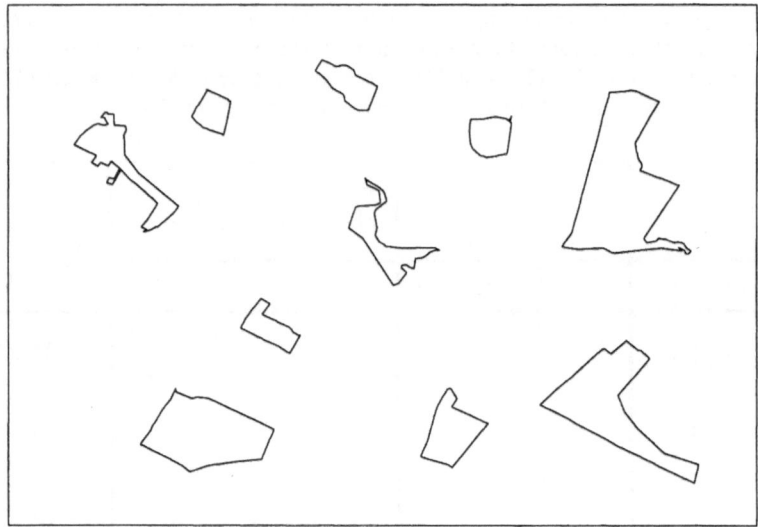

Fig. 7. Ten irregularly shaped polygons from U. S. census geographic data.

A fourth extension is to include complex regions with holes for similar tests and model development. The work presented in this discussion is basically concerned with regions. For instance, future research will need to examine approximate linguistic terms in descriptions involving both linear features and regions. Furthermore, descriptions in other languages in addition to the English language should also be studied.

6   Concluding Remarks

The goal of this chapter was to develop a fuzzy set model of approximate linguistic terms used in natural language descriptions of topological relations between simple crisp regions. This chapter focuses on descriptions of "Region Q coving Region R." People tend to use approximate linguistic terms when describing spatial objects and spatial relations. For instance, it is very common to say: "In this piece of land, there is *basically* soil type M;" or "The flash flood in October of 1998 *nearly completely* flooded downtown San Marcos in Central Texas." In these descriptions, terms such as *'basically'* and *'nearly completely'* are approximate in nature. A formal model that can be used to characterize these approximate linguistic terms is important for developing intelligent computerized systems for processing spatial data and for supporting queries that are described in approximate linguistic terms commonly found in daily communications.

Three approximate linguistic terms, 'a little bit,' 'somewhat,' and 'nearly completely,' may be used to describe the degree of Region Q covering Region R. A good indicator reflecting the degree of Region Q covering Region R is the ratio between the area of the overlapping portion of regions Q and R and the area of Region R itself. This ratio is called the area-based ratio. Preliminary cognitive evidences indicate that: (1) the mean value of the area-based ratio corresponding to Region Q covering Region R 'a little bit' is 0.10, but can vary from 0 to 0.50; (2) the mean value of the area-based ratio corresponding to Region Q covering Region R 'somewhat' is 0.36, but this ratio can vary from 0.15 to 0.70; and (3) the mean value of the area-based ratio corresponding to Region Q covering Region R 'nearly completely' is 0.74, but can vary from 0.65 to 0.99. Based on these cognitive results, a fuzzy set model was developed. This fuzzy set model provides a general formulation for determining fuzzy membership values corresponding to different area-based ratios in relation to the topological relation "Region Q covering Region R." This model, therefore, provides a basis for developing more natural methods for processing topological relations in Geographic Information Systems (GIS).

Acknowledgements

The author wishes to thank the editors of this book for their help in many ways. The author is also grateful to Professor Frederick E. Petry for his comments on an earlier version of this chapter. Assistance from Ms. Ellen Lewis in preparing this chapter is greatly appreciated.

References

1.   R. F. Abler. The National Science Foundation National Center for Geographic Information and Analysis. *International Journal of Geographical Information Systems*, 1(4):303–326, 1987.
2.   D. Altman. Fuzzy set theoretic approaches for handling imprecision in spatial analysis. *Int. Journal of Geographical Information Systems*, 8(3):271-289, 1994.
3.   P. A. Burrough. Fuzzy mathematical methods for soil survey and land evaluation. *Journal of Soil Science*, 40:477-492, 1989.
4.   P. A. Burrough and A. U. Frank (Eds.). *Geographic Objects with Indeterminate Boundaries*, Taylor & Francis, London, 1996.
5.   E. Clementini and P. Di Felice. An algebraic model for spatial objects with indeterminate boundaries. In: P. A. Burrough and A. U. Frank (Eds.), *Geographic Objects with Indeterminate Boundaries*, Taylor & Francis, London, pages 155-169, 1996.
6.   E. Clementini and P. Di Felice. A model for representing topological relationships among complex geometric features in spatial databases. *Information Sciences*, 90:121-136, 1996.

7.   E. Clementini and P. Di Felice. Approximate topological relations. *International Journal of Approximate Reasoning*, 16:173-204, 1997.
8.   A. G. Cohn and N. M. Gotts. The 'Egg-Yolk' representation of regions with indeterminate boundaries. In: P. A. Burrough and A. U. Frank (Eds.), *Geographic Objects with Indeterminate Boundaries*, Taylor & Francis, London, pages 171-187, 1996.
9.   D. Dubois and M.-C. Jaulent. A general approach to parameter evaluation in fuzzy digital pictures. *Pattern Recognition Letters*, 6:251-259, 1987.
10. M. J. Egenhofer and R. Franzosa. Point-set topological spatial relations. *International Journal of Geographical Information Systems*, 5(2):161-174, 1991.
11. M. J. Egenhofer, J. Glasgow, O. Gunther, J. Herring, and D. Peuquet. Progress in Computational Methods for Representing Geographic Concepts. *International Journal of Geographical Information Science*, 13(8):775-796, 1999.
12. M. J. Egenhofer and J. R. Herring. Categorizing Topological Spatial Relations Between Point, Line, and Area Objects. In: M. J. Egenhofer, D. M. Mark, and J. R. Herring, *The 9-Intersection: Formalism and its Use For Natural-Language Spatial Predicates*, Santa Barbara, CA: National Center for Geographic Information and Analysis, Report 94-1, 1994.
13. M. J. Egenhofer and D. M. Mark. Modeling conceptual neighborhoods of topological relations. *Int. Journal of Geographical Information Systems*, 9(5):555-565, 1995.
14. M. Erwig and M. Schneider. Vague Regions. In *SSD'97 (5th Int. Symp. on Advances in Spatial Databases)*, LNCS 1262, pages 298-320, 1997.
15. P. F. Fisher. First experiments in viewshed uncertainty: The accuracy of the viewable area. *Photogrammetric Engineering and Remote Sensing*, 57:1321-1327, 1991.
16. P. F. Fisher. First experiments in viewshed uncertainty: Simulating the fuzzy viewshed. *Photogrammetric Engineering and Remote Sensing*, 58:345-352, 1992.
17. P. F. Fisher. Algorithm and implementation uncertainty in the viewshed function. *International Journal of Geographical Information Systems*, 7:331-347, 1993.
18. J. Freeman. The modeling of spatial relations. *Computer Graphics and Image Processing*, 4:156-171, 1975.
19. N. W. Hazelton, L. Bennett, and J. Masel. Topological structures for 4-dimensional geographic information systems. *Computers, Environment, and Urban Systems*, 16(3):227-237, 1992.
20. G. J. Klir and B. Yuan. *Fuzzy Sets and Fuzzy Logic: Theory and Applications*, Prentice Hall, Upper Saddle River, NJ, 1995.
21. R. Krishnapuram, J. M. Keller, and Y. Ma. Quantitative analysis of properties and spatial relations of fuzzy image regions. *IEEE Transactions on Fuzzy Systems*, 1(3):222-233, 1993.
22. B. Landau and R. Jackendoff. 'What' and 'where' in spatial language and spatial cognition. *Behavioral and Brain Sciences*, 16:217–265, 1993.
23. Y. Leung. Approximate characterization of some fundamental concepts of spatial analysis. *Geographical Analysis*, 14(1):29-40, 1982.
24. Y. Leung. On the imprecision of boundaries. *Geographical Analysis*, 19:125-151, 1987.
25. D. M. Mark. *Languages of Spatial Relations: Researchable Questions and NCGIA Research Agenda*, Santa Barbara, CA: National Center for Geographic Information and Analysis, Report 89-2A, 1989.
26. D. M. Mark. Spatial Representation: A Cognitive View. In: D. J. Maguire, M. F. Goodchild, D. W. Rhind, and P. Longley (Eds.), *Geographical Information Systems: Principles and Applications*, Second edition, John Wiley & Sons, NY, 1:81-89, 1999.
27. D. M. Mark and M. J. Egenhofer. Modeling spatial relations between lines and regions: combining formal mathematical models and human subjects testing. *Cartography and Geographic Information Systems*, 21(4):195–212, 1994.

28. D. M. Mark and M. J. Egenhofer. Topology of Prototypical Spatial Relations Between Lines and Regions in English and Spanish. In *Auto Carto 12 (12th International Symposium on Computer-Assisted Cartography)*, pages 245-254, Charlotte, North Carolina, March 1995.

29. D. M. Mark and A. U. Frank (Eds.). *Cognitive and Linguistic Aspects of Geographic Space*, Kluwer Academic Publishers, Dordrecht, 1991.

30. D. M. Mark, C. Freksa, S. C. Hirtle, R. Lloyd, and B. Tversky. Cognitive Models of Geographic Space. *International Journal of Geographic Information Science*, 13(8):747-774, 1999.

31. P. Matsakis, J. Keller, L. Wendling, J. Marjamaa, and O. Sjahputera. Linguistic Description of Relative Positions in Images. *TSMC Part B (IEEE Trans. on Systems, Man and Cybernetics)*, 31(4):573-588, 2001.

32. NCGIA (National Center for Geographic Information and Analysis). The research plan of the National Center for Geographic Information and Analysis. *International Journal of Geographical Information Systems*, 3(2):117-136, 1989.

33. S. K. Pal and A. Ghosh. Index of area coverage of fuzzy subsets and object extraction. *Pattern Recognition Letters*, 11:831-841, 1990.

34. D. J. Peuquet. Representations of geographic space: toward a conceptual synthesis. *Annals of the Association of American Geographers*, 78:375–394, 1988.

35. D. J. Peuquet and C.-X. Zhan. An algorithm to determine the directional relationship between arbitrarily-shaped polygons in the plane. *Pattern Recognition*, 20:65-74, 1987.

36. G. Retz-Schmidt. Various views on spatial prepositions. *AI Magazine*, 9:95-105, 1988.

37. V. B. Robinson. Implications of fuzzy set theory for geographic databases. *Computers, Environment, and Urban Systems*, 12:89-98, 1988.

38. V. B. Robinson. Interactive machine acquisition of a fuzzy spatial relation. *Computers and Geosciences*, 16:857-872, 1990.

39. V. B. Robinson, M. Blaze, and D. Thongs. Representation and acquisition of a natural language relation for spatial information retrieval. In *Second International Symposium on Spatial Data Handling*, pages 472–487, Seattle, Washington, 1986.

40. V. B. Robinson and R. Wong. Acquiring approximate representation of some spatial relations. In *Auto Carto 8 (8th International Symposium on Computer-Assisted Cartography)*, pages 604-622, 1987.

41. A. Rosenfeld. Fuzzy digital topology. *Information and Control*, 40:76-87, 1979.

42. A. Rosenfeld. Fuzzy rectangles. *Pattern Recognition Letters*, 11:677-679, 1990.

43. A. Rosenfeld and R. Klette. Degree of adjacency or surroundedness. *Pattern Recognition Letters*, 18(2):169-177, 1985.

44. A. R. Shariff, M. J. Egenhofer, and D. M. Mark. Natural-language spatial relations between linear and areal objects: the topology and metric of English-language terms. *International Journal of Geographical Information Science*, 11(3):215–246, 1998.

45. L. Talmy. How language structures space. In: H. L. Pick Jr. and L. P. Acredolo (Eds.), *Spatial Orientation: Theory, Research and Application*, New York, Plenum, pages 225–282, 1983.

46. H. A. Taylor and B. Tversky. Spatial mental models derived from survey and route descriptions. *Journal of Memory and Language*, 31:261–282, 1992.

47. H. A. Taylor and B. Tversky. Perspective in spatial descriptions. *Journal of Memory and Language*, 35:371–391, 1996.

48. D. J. Unwin. Geographical information systems and the problem of 'error and uncertainty.' *Progress in Human Geography*, 19(4):549-558, 1995.

49. F. Wang and G. B. Hall. Fuzzy representation of geographical boundaries in GIS. *International Journal of Geographical Information Systems*, 10(5):573-590, 1996.

50. F. Wang, G. B. Hall, and Subaryono. Fuzzy information representation and processing in conventional GIS software: database design and application. *International Journal of Geographical Information Systems*, 4:261-283, 1990.

51. L. A. Zadeh. Fuzzy Sets. *Information and Control*, 8:338-353, 1965.

52. F. B. Zhan. Approximation of Topological Relations Between Fuzzy Regions Satisfying a Linguistically Described Query (extended abstract). In: S. C. Hirtle and A. U. Frank (Eds.), *COSIT'97, Spatial Information Theory: A Theoretical Basis for GIS*, LNCS 1329, pages 509-510, Laurel Highlands, Pennsylvania, USA, October 1997.

53. F. B. Zhan. Approximate analysis of topological relations between geographic regions with indeterminate boundaries. *Soft Computing*, 2(2):28-34, 1998.

54. F. B. Zhan. How Much is Region Q Covering Region R 'a Little Bit,' 'Somewhat,' or 'Nearly Completely?' In: M. Cristani and B. Bennett (Eds.), *SVUG01: The First COSIT (Conference on Spatial Information Theory) Workshop on Spatial Vagueness, Uncertainty and Granularity*, Morro Bay, CA, September 2001.

About the Editors

Dr. Pascal Matsakis is an Assistant Professor of Computer Engineering and Computer Science at the University of Missouri, Columbia, U.S.A. From Paul Sabatier University, Toulouse, France, he received the B.Sc. degree both in Mathematics and Computer Science, and the Ph.D. in Computer Science (1998). Before moving to the United States, he worked in the research group of Image Processing and Understanding at the Toulouse Institute of Research in Computer Science. His interests include computer vision, computer graphics, human-machine interaction, fuzzy set theory and fuzzy logic. From a general point of view, his research concerns the exploitation of expert knowledge and ancillary data for image interpretation and scene understanding. Dr. Matsakis' particular field is in modeling and utilizing the spatial relations between multidimensional objects.

Dr. Les M. Sztandera is an Associate Professor of Computer Science and Head of the Computer Science Program at the Philadelphia University, Philadelphia, Pennsylvania, U.S.A. He has been involved in soft computing teaching and research since 1987. Dr. Sztandera has 11 years of full time university teaching experience, and is a recipient of a Teaching Excellence Award. He developed a sequence of soft computing courses coupled with laboratory assignments in which students work with real life problems, such as detecting an industrial pollutant, predicting strength and density of materials, designing a medical expert system, simulating protective systems in complex power generating units, detecting carcinogenic dyes, or designing new drugs. Complementary with his teaching effort, Dr. Sztandera has been involved in a variety of research activities. That has resulted in numerous research grants from the Department of Commerce, National Textile Center, National Science Foundation, Ohio Supercomputer Center, Pittsburgh Supercomputer Center, and American Heart Association. Over $1,000,000 in research funding has been experienced. Those research activities also resulted in 25 journal publications and 40 conference presentations.

Dr. Sztandera received his Ph.D. degree from the Department of Electrical Engineering and Computer Science, University of Toledo, Ohio, U.S.A., with a dissertation on Fuzzy Sets in Self-Generating Neural Network Architectures. He earned his M.Sc. degree from the Department of Computer Science and Engineering, University of Missouri, Missouri, U.S.A., with a thesis on Spatial Relations Among Fuzzy Subsets of an Image, and a Diploma in English from University of Cambridge, England.

Dr. Sztandera is a member of professional organizations in the U.S. and Canada: the North American Fuzzy Information Processing Society, Association for Computing Machinery, and Canadian Society for Fuzzy Information and Neural Systems. His scientific and scholarly research contributions to the fuzzy set theory are internationally recognized. He proposed, designed, and implemented fuzzy neural trees. For this and other contributions to the fuzzy sets and systems theory, he was included in the Encyclopedia of Computer Science and Technology, 1999 Edition. Dr. Sztandera is also listed in the Marquis Who's Who in the World, Who's Who in Science and Engineering, Who's Who in America, and Who's Who in the East.